Characterizing the Safety of Self-Driving Vehicles

Characterizing the Safety of Self-Driving Vehicles

JUAN R. PIMENTEL

Professor of Computer Engineering
Kettering University

SAE INTERNATIONAL

Warrendale, Pennsylvania, USA

400 Commonwealth Drive
Warrendale, PA 15096-0001 USA
E-mail: CustomerService@sae.org
Phone: 877-606-7323 (inside USA and Canada)
Fax: 776-4970 (outside USA)

Library of Congress Catalog Number 2019931389
SAE Order Number PT-203
http://dx.doi.org/10.4271/pt-203

ISBN-Print 978-0-7680-0201-0
ISBN-MediaTech 978-0-7680-0210-2
ISBN-prc 978-0-7680-0212-6
ISBN-epub 978-0-7680-0213-3
ISBN-HTML 978-0-7680-0214-0

To purchase bulk quantities, please contact: SAE Customer Service

E-mail: CustomerService@sae.org
Phone: 877-606-7323 (inside USA and Canada)
Fax: 776-4970 (outside USA)

Visit the SAE International Bookstore at books.sae.org

contents

CHAPTER 3

A Model-Driven Approach for Dependent Failure Analysis in Consideration of Multicore Processors Using Modified EAST-ADL   35

CHAPTER 4

An Analysis of ISO 26262: Machine Learning and Safety in Automotive Software   47

SOTIF PAPERS

CHAPTER 5

Hazard Analysis and Risk Assessment beyond ISO 26262: Management of Complexity via Restructuring of Risk-Generating Process   61

MULTI-AGENT SAFETY PAPERS

CHAPTER 8

A Lane-Changing Decision-Making Method for Intelligent Vehicle Based on Acceleration Field   105

Introduction

Safety has been ranked as the number one concern for the acceptance and adoption of autonomous vehicles (AVs) and understandably so since safety has some of the most complex requirements in the development of self-driving vehicles. Recent fatal accidents involving self-driving vehicles have uncovered a host of major issues in the way some automated vehicle companies approach their design, testing, verification, and validation of their products. Traditionally, automotive safety follows functional safety concepts as detailed in the standard ISO 26262. However, autonomous driving safety goes beyond ISO 26262 and includes other safety concepts such as safety of the intended functionality (SOTIF) and multi-agent safety. In this book, we characterize the concept of safety for self-driving vehicles, and we discuss some pertinent literature.

Index Terms

Automated vehicles

Self-driving vehicles

Autonomous vehicles

Safety

SOTIF

Multi-agent safety

Functional safety

I. Introduction

Autonomous vehicles (AVs), also called automated or self-driving vehicles, have the potential to reduce accidents, help with the environment, reduce congestion, help the elderly and other disadvantaged populations, and produce other societal benefits. However, the touted advantages of AVs and those including latest advanced driver assistance system (ADAS) features are turning out difficult to sell to the American public than many manufacturers and tier 1s have anticipated. Much of the early euphoria of self-driving vehicles is diminishing in the wake of some recent accidents involving automated vehicles with varying degrees of automation. A recent online marketplace for buying and selling cars found 69% of respondents are scared of autonomous automobiles. It also found that these people found technology in cars helpful (58%), but only 12% said ADAS and infotainment features were "must have." The survey asked more than 1,000 respondents from across the United States geographically and across age groups, although it should be noted the biggest group was 60+ years old.* Accidents involving self-driving vehicles are inevitable; as it is the case with other industries, accidents have happened and will happen no matter the efforts made to avoid them. The National Transportation Safety Board (NTSB) has issued a report[†] on a Tesla accident on May 7, 2016, and two preliminary reports on an Uber accident[‡] on March 18, 2018, and a Tesla accident[§] on March 23, 2018. After analyzing these reports, what is disturbing are the details associated with these accidents which indicate that as an industry, we may need to go back to safety 101.[¶]

Regarding the NTSB Tesla accident report, on May 7, 2016, a 2015 Tesla Model S 70D car, traveling eastbound on US Highway 27A, west of Williston, Florida, struck a semitrailer resulting in the death of the car driver. System performance data downloaded from the Tesla indicated that the driver was operating the car using features of its autopilot suite: traffic-aware cruise control (TACC) and the autosteer lane-keeping system. The car was also equipped with a forward collision warning (FCW) system and automatic emergency braking (AEB), but those systems did not activate. The NTSB determined that the probable cause of the crash was the truck driver's failure to yield the right of way to the car, combined with the car driver's inattention due to overreliance on vehicle automation, which resulted in the car driver's lack of reaction to the presence of the truck. Contributing to the car driver's overreliance on the vehicle automation was its operational design, which permitted

---

*http://analysis.tu-auto.com/autonomous-car/shifting-public-acceptance-autonomous-tech?NL=TU-001&Issue=TU-001_20180723_TU-001_235&sfvc4enews=42&cl=article_2_2.
† https://www.ntsb.gov/investigations/AccidentReports/Reports/HAR1702.pdf.
‡ https://www.ntsb.gov/investigations/AccidentReports/Reports/HWY18MH010-prelim.pdf.
§ https://ntsb.gov/investigations/AccidentReports/Reports/HWY18FH011-preliminary.pdf.
¶ https://www.eetimes.com/document.asp?doc_id=1333446&_mc=RSS_EET_EDT&utm_source=newsletter&utm_campaign=link&utm_medium=EETimesWeekInReview-20180721.

his prolonged disengagement from the driving task and his use of the automation in ways inconsistent with guidance and warnings from the manufacturer.

On the NTSB Uber accident preliminary report, on March 18, 2018, an Uber test vehicle, based on a modified 2017 Volvo XC90 and operating with a self-driving system in computer control mode, struck and killed a pedestrian on northbound Mill Avenue, in Tempe, Arizona. According to data obtained from the self-driving system, the system first registered radar and LIDAR observations of the pedestrian about 6 seconds before impact, when the vehicle was traveling at 43 mph. As the vehicle and pedestrian paths converged, the self-driving system software classified the pedestrian as an unknown object, as a vehicle, and then as a bicycle with varying expectations of future travel path. At 1.3 seconds before impact, the self-driving system determined that an emergency braking maneuver was needed to mitigate a collision. According to Uber, emergency braking maneuvers are not enabled while the vehicle is under computer control, to reduce the potential for erratic vehicle behavior. The vehicle operator is relied on to intervene and take action. The system was not designed to alert the operator.

On the NTSB Tesla accident preliminary report, on March 23, 2018, a 2017 Tesla Model X P100D vehicle was traveling south on US Highway 101 in Mountain View, California. As the vehicle approached the US-101/SH-85 interchange, it was traveling in the second lane from the left, which was a high-occupancy vehicle (HOV) lane for continued travel on US-101. According to performance data downloaded from the vehicle, the driver was using the advanced driver assistance features TACC and autosteer lane-keeping assistance, which Tesla refers to as "autopilot." As the Tesla approached the paved gore area dividing the main travel lanes of US-101 from the SH-85 exit ramp, it moved to the left and entered the gore area. The Tesla continued traveling through the gore area and struck a previously damaged crash attenuator at a speed of about 71 mph. The TACC speed was set to 75 mph at the time of the crash. The Tesla was involved in subsequent collisions with two other vehicles, a 2010 Mazda 3 and a 2017 Audi A4. The Tesla driver was taken to a hospital and later died from his injuries.

Regarding the aforementioned accidents, it would not be so bad if the safety systems of the vehicles in question were designed and functioning properly according to their stated automation level. In the case of the Tesla accident in Florida, the vehicle failed to activate the FCW system and AEB. In the case of the Uber accident, emergency braking maneuvers were not enabled while the vehicle was under computer control, to reduce the potential for erratic vehicle behavior, and the system relied on the vehicle operator for safety. In the Tesla accident in California, the vehicle failed to detect a damaged crash attenuator and hit it at a speed of about 71 mph.

After recent incidents and mishaps involving automated vehicles such as those described above, it is clear that there is much room for improvement not only by manufacturers but also by government regulations, researchers, the general public, and other stakeholders. Over the past few months, the media has been full of headlines such as "How

Safe Is Driverless Car Technology, Really?", "Autonomous Cars: How Safe Is Safe Enough?", and "How safe should we expect self-driving cars to be?". In addition, some industry analysts and safety experts are offering advice to tech and automotive companies to reconsider their safety programs. There is also some agreement that "the self-driving car industry's reputation has suffered a setback," and the question is how to fix it*. It appears that automated vehicle companies are much more stringent when using semiconductor devices and EDA tools demanding that they conform to ISO 26262 than using the same yardstick for their own safety-critical designs.

So what is there to do? Safety is not new, at least for the last 60 years it has been successfully applied in several industries such as nuclear, avionics, process control, automotive, and others. What is unique and special about the safety of self-driving vehicles? What should be the emphasis for a more effective automated vehicle safety program? What are the roles of governments, standards, testing, verification, validation, and sound safety engineering efforts? As an industry, we just do not fully understand the nature of self-driving vehicle safety and how to design safe automated vehicles. For example, there are little discussions on ways to estimate, analyze, compute, or measure the level of safety of an automated vehicle design. We need to begin by fully characterizing it and this book is an effort in this direction.

Some manufacturers such as Waymo cite their recent milestone of 8 million miles driven on public roads as a measure of the safety achieved by their self-driving vehicles.† However, it is not clear how a certain number of millions of miles driven contributes to the safety level of self-driving vehicles. Some industry analysts believe that policy makers and city officials overseeing infrastructure will be the most important players in reshaping the self-driving vehicle safety landscape. For example, NHTSA has issued a voluntary guidance whose purpose is to help designers of automated driving systems (ADSs) analyze, identify, and resolve safety considerations prior to deployment using their own, industry, and other best practices.‡ It outlines 12 safety elements, which the agency believes represent the consensus across the industry, that are generally considered to be the most salient design aspects to consider and address when developing, testing, and deploying ADSs on public roadways. Within each safety design element, entities are encouraged to consider and document their use of industry standards, best practices, company policies, or other methods they have employed to provide for increased system safety in real-world conditions. The 12 safety design elements apply to both ADS original equipment and to replacement equipment or updates (including software updates/upgrades) to ADSs.

---

* https://www.eetimes.com/document.asp?doc_id=1333446&_mc=RSS_EET_EDT&utm_source=newsletter&utm_campaign=link&utm_medium=EETimesWeekInReview-20180721.
† https://www.theverge.com/2018/7/20/17595968/waymo-self-driving-cars-8-million-miles-testing.
‡ https://www.nhtsa.gov/sites/nhtsa.dot.gov/files/documents/13069a-ads2.0_090617_v9a_tag.pdf.

However, the NHTSA guidance is not specific enough to help manufacturers in designing effective safety mechanisms to reduce risk.

II. Brief Overview of Self-Driving Vehicles

The main components of an AV include a perception system, a localization system, a mission and path planning system, a control system, and an actuation system. Over the years, there has been much progress in developing full self-driving vehicles by manufacturers and academics [1, 2, 3, 4, 5]. Nevertheless, one of the current outstanding issues is how to develop AVs with an adequate level of safety, such that they can be accepted by the majority of the population at a reasonable cost. The recent fatal accidents involving self-driving vehicles have made it clear that safety is paramount to the acceptance, testing, verification, and deployment of AVs.

The perception and control systems are perhaps the most safety-critical components of an AV, because of their complexity and the vehicle's potential to produce harm, mishaps, or accidents. Thus, the perception and control systems of AVs must be designed with the utmost care using state-of-the-art methodologies from the safety engineering field. Addressing safety issues in engineering has a long history that began with the nuclear industry, then avionics, and then followed by other industries such as process control and automotive. Although the fundamental safety concepts are the same in all these industries, the safety characterization and application to a particular industry is more specific. Because of its complexity, we do not currently fully understand the concept of the safety of self-driving vehicles. AVs pose perhaps the greatest of challenges to safety, because of the complexities of their perception and control systems that need to operate in real time in a highly dynamic environment. This environment includes other vehicles that may be traveling at high speeds. Safety has been ranked as the number one concern for the acceptance and adoption of AVs and understandably so since safety has some of the most complex requirements in the development of self-driving vehicles. Traditionally, automotive safety follows functional safety concepts as detailed in the standard ISO 26262 [6]. However, autonomous driving safety goes beyond ISO 26262 and includes other safety concepts such as safety of the intended functionality (SOTIF) and multi-agent safety. In addition, governments at all levels are stepping in to help define and address multiple safety requirements appropriate for AVs. Thus, safety has become the most important concern for automakers, service providers, governments, and all associated stakeholders in the self-driving ecosystem.

A deep understanding of AV safety is difficult to acquire all by itself. In addition to safety knowledge acquisition and creation, however, there is also the need of a holistic solution that will necessarily include a broad appreciation for the range of challenges (and potential solutions) by all of the relevant disciplines. As Koopman and Wagner pointed out:

> *Safe means at least correctly implementing vehicle- level behaviors such as obeying traffic laws (which can vary depending upon location) and dealing with nonroutine road hazards such as downed power lines and flooding. But it also means things such as fail-over mission planning, finding a way to validate inductive- based learning strategies, providing resilience in the face of likely gaps in early-deployed system requirements, and having an appropriate safety certification strategy to demonstrate that a sufficient level of safety has actually been achieved. [7]*

Safety, one of the attributes of dependability, is defined in [8] as the absence of catastrophic consequences (e.g., accidents or harm) for the user(s) and the environment. Failures can be catastrophic or non-catastrophic (also called benign). The state of correct service and the state of incorrect service due to non-catastrophic failures are grouped into a safe state, indicating freedom from catastrophic damage. Safety is a measure of continuous safeness, or, equivalently, it is a measure of the time to catastrophic failure. A system failure is an event that occurs when the delivered service deviates from the correct service. A failure, therefore, is a transition from correct service to incorrect service, that is, to non-implementation of the system function. ISO 26262 defines a failure as the termination of the ability of an element or system in the performance of its required function. Thus, the definition in [8] is more generic than that in ISO 26262, and it includes services such as those provided by the perception system of a self-driving vehicle. Safety is characterized, analyzed, and addressed using the concepts of safety hazard and safety risk.

As emphasized in [7], ensuring the safety of fully AVs requires a multi-disciplinary approach across all the levels of functional hierarchy, from hardware fault tolerance to resilient machine learning (ML), to cooperating with humans driving conventional vehicles, to validating systems for operation in highly unstructured environments, to appropriate regulatory approaches. Significant open technical challenges include validating inductive learning in the face of novel environmental inputs and achieving the very high levels of dependability required for full-scale fleet deployment. Significant challenges must be overcome to achieve self-driving vehicles with a sufficient level of safety, not only in safety engineering but in other areas such as computing hardware, software, robotics, security, testing, human-computer interaction, social acceptance, and law [7].

A. Autonomous Driving Model

We view the overall mission of a self-driving vehicle as composed of a hierarchy of tasks, starting with the top-level mission that can be called a trip, which can be further decomposed into several trip sections.

TABLE 1 Relationship among functionalities, states, events, behaviors, and CAs

Functionality	Follow a road	Handling a cut-in on highway
States	Following the road at normal speed, following the road while slowing down, following the road while accelerating	Changing to left lane, changing to right lane, continuing on the same lane
Event	Stop sign detected	Vehicle detected
Behavior	Slow down while following a road	Turn left to avoid hitting the detected vehicle
CAs	Brake, accelerate	Turn left, turn right, brake

For example, a typical trip from home to work might involve the following trip sections: leaving the garage, driving near home, driving through the neighborhood, driving on the highway, driving through the city, entering a parking lot, and parking. We view each trip section as a combination of a set of self-driving functionalities that involve road detection, lane detection, obstacle detection, stop sign detection, vehicle detection, pedestrian detection, cyclist detection, traffic light detection, highway entrance negotiation, highway exit negotiation, road merging, highway merging, etc. Each self-driving functionality involves several concepts that are useful for designing safe self-driving vehicles, which include intended functionality or simply functionality, state, event, behavior, and control action (CA).

While functionality can be defined as the quality or capacity of performing functions, behavior involves actions required to provide a specific function or functions. A state is a mode of being of a system or sub-system while performing a function. An event can be anything of importance that takes place or happens that has the potential to change states. A CA is something that is done with the potential to generate events or change states of the system. We distinguish several types of CAs: an emergency control action (ECA), a safe control action (SCA), and an unsafe control action (UCA). From this perspective, the terms functionality and behavior can be used interchangeably. As an example, Table 1 lists two functionalities with their respective states, events, behaviors, and CAs.

In this book, we address some specific functionalities in detail that involve vehicle detection. Table 2 lists examples of states, SCAs, and UCAs pertaining to a vehicle detection event.

The goal of this book is to characterize safety in the context of AVs.

It is hoped that this introductory chapter contributes to the characterization of safety for AVs that involves a new framework together with new concepts and definitions.

The existing state of the art includes an emphasis on the use of control and systems engineering concepts for safety and risk analysis, approaches based on systems-theoretic process analysis (STPA) for safety verification, development of test cases, hazard identification, hazard classification, and an incipient research on fault-tolerant designs for self-driving vehicles.

TABLE 2 Some states, safe control actions, and unsafe control actions for a vehicle detection event

Event	State	SCA	UCA
Vehicle detection	Waiting to turn left on a street or road with a stop sign (no traffic signal)	Turn left if distance to vehicle > minimum distance	Turn left if distance to vehicle < minimum distance
	Waiting at a traffic light and light turns green	Wait until vehicle clears	Drive ahead
	Moving on a lane	Maintain a safe distance to the detected vehicle in front	Fail to maintain a safe distance to the detected vehicle in front
	Moving on a lane and detecting a vehicle cut-in	Perform a safe cut-in maneuver	Fail to perform a safe cut-in maneuver
	Attempting to merge into a highway	Slow down and allow vehicles to pass	Accelerate and merge
	Exiting garage	Stop	Do not stop

SCA, safe control action; UCA, unsafe control action

III. Self-Driving Vehicle Safety Attributes

How different is the concept or notion of safety in self-driving vehicles when compared to that used in other industries such as aviation, process control, and automotive? While the fundamental concepts are the same, the safety of self-driving vehicles has specific attributes that are different or not present in the safety of other industries. In this section, we briefly discuss these attributes. When compared to the safety of traditional industries such as avionics, process control, and automotive, there are specific attributes pertaining to the safety of self-driving vehicles that we discuss next.

A. Performance Degradation

Traditional safety is based on faults and failures of mostly hardware components, and this is referred to as the reliability approach to safety [9]. In contrast, accidents involving self-driving vehicles might happen even if no hardware device fails, but rather a performance degradation of some of its functions or intended functionality occurs. Addressing safety issues for these situations is referred to as SOTIF, and it is a fairly new concept as applied to the safety of self-driving vehicles [10]. Thus, some failures are due to performance degradation of self-driving vehicle components, typically involving higher levels of processing or higher levels of automation, for example, service failures. This definition of failure goes beyond that which is defined in the standard ISO 26262; however, it is compatible with other safety frameworks such as [8], STPA [11, 12, 13, 14, 15], or other real-time distributed systems [16]. One example of this understanding of the concept of safety is the failure of a vehicle detection system where the perception system provides missed detections (i.e., false negatives) or spurious detections (i.e., false positives).

This could happen because the processing of environmental data is highly complex and the object detection function is subject to errors and impairments, particularly in bad weather or in night conditions when the visibility is poor. Either one of these failures could be catastrophic and could result in an accident or harm. Another example is a radio detection and ranging (RADAR) system correctly detecting objects only when the objects are moving, thus missing static objects because of limits on its performance. Thus, failure occurs in a degraded performance scenario.

B. Focus on Software

It is well known that the amount of software in a vehicle continues a rapidly increasing trend that started with the development of by-wire systems. Much of the functionality of a self-driving vehicle is implemented in software, and thus it is important to view the perception system as a set of software servers each providing services to the rest of the system. Therefore, one can refer to these various functionalities as a vehicle detection server, a pedestrian detection server, a road detection server, etc. The software in self-driving vehicles is much larger in size and scope compared to traditional industries; thus, there should be a focus on the safety of the software. As noted, a failure can occur if the software services deviate from the correct services, and this could lead to safety hazards and safety risks. Ultimately, the overall safety of a self-driving vehicle will be dictated by the safety of its software [9, 12].

C. Nondeterministic Perception System

In the absence of hardware faults, the perception systems of traditional industries are mostly deterministic in nature. For example, sensing the intake manifold pressure or engine speed in automotive systems is deterministic. In contrast, the perception systems of self-driving vehicles are nondeterministic, leading to a high level of false positives and false negatives when their performance deteriorates to the point that service failures cannot be avoided. The nondeterministic aspect of the perception system stems from the fact that one never knows when its performance will deteriorate to the point where failures begin to appear in the services delivered by the system. Thus, the services provided by the perception system are subject to random failures, for example, when the weather deteriorates or when the system makes detection errors.

D. Perception System Complexity

Sensing elementary physical phenomena such as temperature or pressure is relatively simple, involving just some deterministic sensors, some electronics, and communications. In contrast, sensing or detecting man-made entities or constructs such as another vehicle, a road boundary, a city street, or a street intersection is complex because of the lack of structure of what is being sensed or perceived.

The implication of the complexity of the perception system is that it is prone to errors, which degrades the performance or safety of the overall vehicle.

E. Overall System Complexity

In addition to its perception system, a self-driving vehicle also includes localization and mapping, planning and control, and actuation resulting in a highly complex system. One of the main issues with system complexity is that it makes testing for safety challenging, particularly if ML techniques are used, as it makes the design opaque to humans. This makes tracing the design and the test plans to the requirements problematic, since there is no human-understandable design that can be used for verification and testing [17]. In addition, it is known that when the system is complex, the system safety is affected by interacting complexity and tight coupling [9]. Another aspect of system complexity is that the AV operates in a complex external environment, and there are safety hazards due to events outside the domain of the AV, for example, from other vehicles (whether self-driving or not). Thus, the safety attributes of a self-driving vehicle are significantly different from those in other industries such as avionics, process control, and automotive.

IV. Characterizing the Safety of Self-Driving Vehicles

In addition to its attributes, what are the various types of safety that encompass the overall safety of self-driving vehicles? As noted, the safety of self-driving vehicles is complex and differs from that of other industries such as avionics, process control, and automotive. On the one hand, there are safety commonalities such as the safety that involves component failures (CFs), which is the subject of so-called functional safety, and the safety involving components whose failure rates are well understood because they are proven in use, that is, in actual operation. On the other hand, there are two types of safety that are not prevalent in the avionics, process control, and automotive industries, and these include SOTIF and multi-agent safety.

Thus, the types of safety that characterize the safety of self-driving vehicles include (1) traditional functional safety as defined by ISO 26262, (2) SOTIF, and (3) multi-agent safety, which together we refer to as operational safety. Feth et al. also emphasize that safety assurance is a concern because established safety engineering standards and methodologies are currently not sufficient [22]. They also conclude that there are three types of safety that characterize the safety of self-driving vehicles: (1) traditional functional safety, (2) SOTIF which they assume is due to functional insufficiencies, and (3) multi-agent safety related to safe driving behaviors which are abstracted from technological challenges of situation awareness. Furthermore, they elaborate the fundamental safety engineering steps that are necessary to create safe vehicle of higher automation levels while mapping these steps to the guidance presently available in existing (e.g., ISO26262) and upcoming (e.g., ISO PAS 21448) standards.

Functional safety is a well-understood area which is guided by a number of international standards such as IEC 61508 [18], IEC 61511 [19], and ISO 26262, and there is a large number of papers and publications on this topic. However, it is noted that ISO 26262 does not cover automated vehicles; thus its application should be done with great care. In the following, we summarize the safety categories known as SOTIF and multi-agent safety.

A. Safety of the Intended Functionality (SOTIF)

SOTIF is related to the performance degradation attribute discussed in section II. The standard ISO 26262 defines intended functionality as the behavior specified for an item, component, or service excluding safety mechanisms. This concept is similar to that of the Basic Process Control System (BPCS) in IEC 61511, which also includes just control functions, excluding safety. Thus, all functions of the perception system such as vehicle detection, lane detection, road detection, and pedestrian detection are intended functionalities according to ISO 26262. As noted in section II, it is possible that accidents can happen even if all of the hardware components (i.e., sensors, computing units, communication networks) of the perception system are fault free, due to errors in the sensor processing and perception functions (e.g., vehicle detection). The safety that is associated with failures in the services provided by the perception system functionality, which are caused by performance degradation rather than random hardware or software failures, is referred to as SOTIF.

SOTIF is a fairly new concept that encompasses the notion of assuring the safety of systems that contain intended functionalities; much more research is needed to fully characterize it [10]. Thus, a fundamental question involving the SOTIF is how to design systems with an acceptable level of safety, even when sub-systems providing intended functionalities are prone to performance-related failures. It is then important to reduce risk caused by failures in the intended functionality using proper techniques and methodologies, for example, using fault-tolerant designs. What is interesting is that much of the discussion about SOTIF is used in the context of the perception system, for example, detecting a vehicle. However, SOTIF is not restricted to perception, as it is applicable to any function with degraded functionality. Thus, the actual meaning of SOTIF refers to the safety measures or mechanisms of applications that involve functions, with performance that can be degraded to the point where it constitutes a safety hazard, such as vehicle or pedestrian detection.

B. Multi-Agent Safety

In terms of safety, designing an autonomous flying machine is much easier than designing a self-driving vehicle, because currently there is simply much less in the sky to deal with in regard to other flying machines or agents, in general. Unlike avionics, process control, and automotive, accidents in AVs can happen due to actions of agents such as other vehicles, pedestrians, bicyclists, and in general any object that is external to the vehicle. Addressing safety issues in this type of environment is known as multi-agent safety. An example of a specific multi-agent situation that might involve a safety hazard is that of a cut-in that happens when a vehicle changes lanes and positions itself ahead of the self-driving vehicle on a highway. For this case, the set of CAs are brake, turn left, and turn right, which can lead to accidents if not performed safely. As is the case for SOTIF, much more research is needed to fully characterize multi-agent safety [20, 21].

V. The Papers in This Collection

There are ten papers in this book collection, four in the category of functional safety, two in SOTIF, and four in multi-agent safety.

A. Functional Safety Papers
1. The Development of Safety Cases for an Autonomous Vehicle: A Comparative Study on Different Methods, by Junfeng Yang, Michael Ward, and Jahangir Akhtar
2. A Means of Assessing the Entire Functional Safety Hazard Space, by Daniel Aceituna
3. A Model-Driven Approach for Dependent Failure Analysis in Consideration of Multicore Processors Using Modified EAST-ADL, by Bülent Sari and Hans-Christian Reuss
4. An Analysis of ISO 26262: Machine Learning and Safety in Automotive Software, by Rick Salay, Rodrigo Queiroz, and Krzysztof Czarnecki

B. SOTIF Papers
1. Hazard Analysis and Risk Assessment Beyond ISO 26262: Management of Complexity via Parametrization, by Oleg Lurie and Joseph Miller
2. The Science of Testing: An Automotive Perspective, by Siddartha Khastgir, Stewart Birrell, Gunwant Dhadyalla, and Paul Jennings

C. Multi-Agent Safety Papers
1. Theory of Collision Avoidance Capability in Automated Driving Technologies, by Toshiki Kindo and Bunyo Okumura
2. A Lane-Changing Decision-Making Method for Intelligent Vehicle Based on Acceleration Field, by Bing Zhu, Shuai Liu, and Jian Zhao

3. Entropy in Reaction Space - Upgrade of Time-to-Collision Quantity, by Vaclav Jirovsky

4. A Maneuver-Based Threat Assessment Strategy for Collision Avoidance, by Yaxin Li, Weiwen Deng, Bohua Sun, Jian Zhao, and Jinsong Wang

In the following, we summarize the above papers in the context of the characterization of the safety of self-driving vehicles. After this introduction, the actual papers follow.

A. Functional Safety Papers

1. The Development of Safety Cases for an Autonomous Vehicle: A Comparative Study on Different Methods, by Junfeng Yang, Michael Ward, and Jahangir Akhtar

   In this paper, the authors argue that to achieve complete safety, a safety case providing guidance on the identification and classification of hazardous events and the minimization of these risks needs to be developed throughout the entire development life cycle process of automated vehicles. It is stated that a comprehensible and valid safety case has to employ appropriate safety approaches complying with the automotive functional safety requirements in ISO 26262. The paper includes a comparative study of different safety approaches, in particular, failure mode and effects analysis (FMEA) method and goal structuring notation (GSN) method that have been employed to generate lists of hazardous events, safety goals, and functional safety requirements at the vehicle level. A case study on the safety case development of INSIGHT AV is included using the aforementioned methods. This case study covers the safety argument of battery and charging system that supply the whole electric power for INSIGHT vehicle. The safety of these sub-systems has been assessed along with their potential for malfunction together with the layers of protection. The results and conclusions from case study analyses suggest the safety case of automated vehicles can be developed in a highly effective manner by employing a combined method of GSN and FMEA. Although the case study is for the battery and charging system of an automated vehicle, the approach is valid for other sub-components such as the perception system.

2. A Means of Assessing the Entire Functional Safety Hazard Space, by Daniel Aceituna

   A major challenge in determining hazard scenarios is trying to assess an adequate amount of scenarios, considering the large size of a hazard space. Typically assessing the entire hazard space is difficult to achieve, resulting in the possibility of overlooking some critical scenarios that can result in harm to either system operators, system bystanders, or both. The paper presents a rule-based approach for concisely describing hazard scenarios, which could potentially enable one to examine the entire hazard space in a short amount of time. The approach, called hazard space analysis (HSA), utilizes a rule-based notation that describes a hazard scenario as a transition from a system state (operational situation (OS)) to a hazard state (HS) as caused by one of several possible actions/ states. The rule-based format also allows for a concise, high-level description of a hazard scenario that maps directly into a safety requirement. Each rule consists of three artifacts, OS, HS, and a hazard cause, which in turn can be classified as either an environmental state (ES), an operator nominal behavior (NB), an operator off-nominal behavior (ONB), or a CF. These various classifications allow for the partitioning of the hazard space, as well as a framework for the potential combining of hazard causes.

Furthermore, HSA combines three key activates: determining hazard scenarios, assigning a risk factor to those scenarios, and mapping those hazard scenarios directly to safety rules. We will detail the approach, show how the approach could be automated, and present a simple aviation-related example that demonstrates the approach's potential in enabling stakeholders to explore a large hazard space. Although the use case described in the paper is in aviation based on IEC 61508, the method could be easily applied to a self-driving vehicle using ISO 26262.

3. A Model-Driven Approach for Dependent Failure Analysis in Consideration of Multicore Processors Using Modified EAST-ADL, by Bülent Sari and Hans-Christian Reuss

Increasing functionality of vehicle systems through electrification of power train and autonomous driving leads to complexity in designing system, hardware, software, and safety architecture. The application of multicore processors in the automotive industry is becoming necessary because of the needs for more processing power, more memory, and higher safety requirements. Therefore, it is necessary to investigate the safety solutions particularly for Automotive Safety Integrity Level D (ASIL-D) systems. This brings additional challenges because of additional requirements of ISO 26262 for ASIL-D safety concepts. This paper presents an approach for model-based "dependent failure analysis" which is required from ISO 26262 for ASIL-D safety concepts with decomposition approach. Therefore, the hardware modeling, function modeling, and dependability package of EAST-ADL (Electronics Architecture and Software Technology – Architecture Description Language) are extended in a way that it now allows the modeling of a multicore processor with its hardware elements and software safety architecture which are necessary to prove hardware and software independency. Additionally, some scripts are developed to analyze the decomposition paths automatically from system level to software and hardware level and generate the analysis results. In addition, the paper describes how a model can be used to develop the main activities of ISO 26262 such as hazard analysis and risk assessment, functional safety concept, technical safety concept, safety analysis, etc. The extensions and developed scripts make it possible to gain sufficient transparency and traceability for the safety arguments and to support the whole safety process in a single solution even during the development of hardware and software.

4. An Analysis of ISO 26262: Machine Learning and Safety in Automotive Software, by Rick Salay, Rodrigo Queiroz, and Krzysztof Czarnecki

ML plays an ever-increasing role in advanced automotive functionality for driver assistance and autonomous operation; however, its adequacy from the perspective of safety certification remains controversial. This paper analyzes the impacts that the use of ML within software has on the ISO 26262 safety life cycle, identifies some issues, and asks what could be done to address them. The authors provide a set of recommendations on how to adapt the standard to better accommodate ML. The most important identified issues are (a) the use of ML can create new types of hazards which can negatively impact safety even though there is no system malfunction or misuse and (b) ML violates the assumption in ISO 26262 that component behavior is fully specified and each refinement can be verified with respect to its specification. This assumption is violated when a training set is used in place of a specification since such a set is necessarily incomplete, and it is not clear how to create assurance that the corresponding hazards are always mitigated.

The most important recommendations for ISO 26262 are (a) the definition of hazard should be broadened to include harm potentially caused by complex behavioral interactions between humans and the vehicle that are not due to a system malfunction; (b) the distinctive types of ML faults create the opportunity to develop focused tools and techniques to help find faults independently of the domain for which the ML model is being trained; and (c) the approach required for high ASIL component implementation should be based on the specifiability of the functionality being implemented. For functionality that is fully specifiable, programming must be required. For functionality that admits no complete specification (e.g., perception), ML-based approaches should be allowable, and the complete specification requirement must be relaxed.

B. SOTIF Papers

1. Hazard Analysis and Risk Assessment Beyond ISO 26262: Management of Complexity via Parametrization, by Oleg Lurie and Joseph Miller

The automotive world is getting ready to embrace the automated driving (AD), while ADAS increases their authority in the control over the vehicle. It is necessary to guarantee system safety of the AD/ADAS application, which includes both "classic" functional safety according to ISO 26262 and specific areas like SOTIF and others. However, safety remains safety, that is, the absence of unreasonable risk. All safety activities within a project, therefore, need to have their source in a hazard analysis and risk assessment (HARA), encompassing all relevant aspects, including OS, description of functionality, and other parameters. Already from the description, a HARA for an AD/ADAS is going to be a complex task. The following features distinguish the HARA related to the SOTIF from the HARA described in ISO 26262: (a) while performing hazard analysis, although the severity and controllability estimations use the same scales, their determination is specific for SOTIF hazards; (b) safety levels are not specifically addressed. The term "acceptable risk," which is often used in ISO 26262, refers to the acceptability of severity and controllability (S0 and C0 evaluations, respectively); and (c) SOTIF HARA includes the specification of a validation target requiring method of validation to be specified. The authors demonstrate an approach for complexity management of HARA for ADAS. The paper provides a manageable overview of potential hazards resulting from malfunctions as well as from external causes including the definition of SOTIF validation goals.

2. The Science of Testing: An Automotive Perspective, by Siddartha Khastgir, Stewart Birrell, Gunwant Dhadyalla, and Paul Jennings

    Increasing automation in the automotive systems has refocused the industry's attention on verification and validation methods and especially on the development of test scenarios. The complex nature of ADAS and ADSs warrant the adoption of new and innovative means of evaluating and establishing the safety of such systems. In this paper, the authors discuss the results from a semi-structured interview study, which involved interviewing ADAS and AD experts across the industry supply chain. Eighteen experts (each with over 10 years of experience in testing and development of automotive systems) from different countries were interviewed on two themes: test methods and test scenarios. Each of the themes had three guiding questions which had some follow-up questions. The interviews were transcribed and a thematic analysis via coding was conducted on the transcripts. A two-stage coding analysis process was done to first identify codes from the transcripts, and, subsequently, the codes were grouped into categories. The analysis of transcripts for the question about the biggest challenge in the area of test methods revealed two specific themes: first, the definition of pass/fail criteria and, second, the quality of requirements (completeness and consistency). The analysis of the questions on test scenarios revealed that "good" scenario is one that is able to test a safety goal and ways in which a system may fail. Based on the analysis of the transcripts, the authors propose two types of testing for ADAS and ADSs: requirements-based testing (traditional method) and hazard-based testing. The proposed approach not only generates test scenarios for testing how the system works but also how the system may fail.

C. Multi-Agent Safety Papers

1. Theory of Collision Avoidance Capability in Automated Driving Technologies, by Toshiki Kindo and Bunyo Okumura

   This paper proposes a theory to analyze the collision avoidance capability of automated driving technologies. The theory gives answers to a fundamental question whether automated vehicles fall into extreme conditions at all rather than another question how a vehicle reacts under extreme conditions (is it as safe as driver?). There are two types of hazards that can cause collisions, cognitive hazards and behavioral hazards. Cognitive hazards are handled by controlling the upper limit speed of the automated vehicle including when stopped. There are two methods for handling behavioral hazards, preparation and response. The response known well is the coping method activated when the hazard is detected in the dynamic (operational) level. The preparation is the coping method operating at all time in the semantic (tactical) level. The collision condition in the semantic level is as follows: a collision occurs when the paths of two vehicles have a crossing point and the two vehicles drive on the crossing point at same time. The condition can be formulated as collision avoidance equation. Solving the equation means that the automated vehicle has prepared for the behavioral hazard before the hazard occurs. It is concluded that a collision avoidance capability consists of not only a response capability that supports the accuracy of collision avoidance in extreme conditions in the dynamic level but also a preparation capability that supports the accuracy to avoid reaching those extreme conditions in the semantic level. The preparation capability can be evaluated through stability analysis of the automated vehicle behavior given by the temporal backward simulation from each extreme condition. A remaining problem is how to determine the upper limit of the hazard's growing speed to which the automated vehicles should react.

2. A Lane-Changing Decision-Making Method for Intelligent Vehicle Based on Acceleration Field, by Bing Zhu, Shuai Liu, and Jian Zhao

   Taking full advantage of available traffic environment information, making control decisions, and then planning trajectory systematically under structured road conditions are a critical part of intelligent vehicle. In this paper, a lane-changing decision-making method for intelligent vehicle is proposed based on acceleration field. Firstly, an acceleration field related to relative velocity and relative distance is built based on the analysis of braking process, and acceleration was taken as an indicator of safety evaluation. Then, a lane-changing decision method is set up with acceleration field while considering driver's habits, traffic efficiency, and safety. Furthermore, velocity regulation is also introduced in the lane-changing decision method to make it more flexible. Afterward, the polynomial trajectory planning method is matched up with this lane-changing decision-making method, and simulations based on Matlab/Simulink are finally conducted to verify the proposed method. Simulation results show that, by adopting the lane-changing decision-making method based on acceleration field, the lane-changing measurements such as starting position, span, and driving speed can be optimized with driver's habits involved. At the same time, vehicle safety can be well ensured.

3. Entropy in Reaction Space - Upgrade of Time-to-Collision Quantity, by Vaclav Jirovsky

Today's vehicles are being more often equipped with systems, which are autonomously influencing the vehicle behavior. More systems of the kind and even fully AVs in regular traffic are expected by OEMs in Europe around year 2025. Driving is a highly multitasking activity, and human errors emerge in situations, when he is unable to process and understand the essential amount of information. Future autonomous systems very often rely on some type of inter-vehicular communication. This shall provide the vehicle with higher amount of information, than the driver uses in his decision-making process. Therefore, currently used 1-D quantity time-to-collision (TTC) will become inadequate. Regardless of whether the vehicle is driven by human or robot, it's always necessary to know whether and which reaction is necessary to perform. Adaptable AV systems will need to analyze the driver's situation awareness level. Such knowledge can be enhanced by 2-D quantity, so-called reaction space, and its entropy. The new approach defines a limit space, where ego vehicle or other vehicles can be present in the future specified by an amount of time. This enables the option of counting not only with braking time, but mitigation by changing direction is feasible. As opposed to TTC, considering time as an input is appreciated especially when switching from autonomous to manual driving. For such situation we observe two kinds of reaction spaces—first, connected with the requirements of autopilot, and, second, resulting from the expected human reaction. Effects of entropy in 2-D reaction space are presented in the paper.

4. A Maneuver-Based Threat Assessment Strategy for Collision Avoidance, by Yaxin Li, Weiwen Deng, Bohua Sun, Jian Zhao, and Jinsong Wang

ADAS are being developed for more and more complicated application scenarios, which often require more predictive strategies with better understanding of driving environment. Taking traffic vehicles' maneuvers into account can greatly expand the beforehand time span for danger awareness. This paper presents a maneuver-based strategy to vehicle collision threat assessment. First, a maneuver-based trajectory prediction model (MTPM) is built, in which near-future trajectories of ego vehicle and traffic vehicles are estimated with the combination of vehicle's maneuvers and kinematic models that correspond to every maneuver. The most probable maneuvers of ego vehicle and each traffic vehicles are modeled and inferred via Hidden Markov Models with mixture of Gaussians outputs (GMHMM). Based on the inferred maneuvers, trajectory sets consisting of vehicles' position and motion states are predicted by kinematic models. Subsequently, TTC is calculated in a strategy of employing collision detection at every predicted trajectory instance. For this purpose, safe areas via bounding boxes are applied on every vehicle, and separating axis theorem (SAT) is applied for collision prediction, so that TTC can be calculated efficiently and accurately. Finally, a threat level index based on reverse TTC is used to quantize the threat degree of every traffic vehicle's potential collision to the ego vehicle. Experimental data collected in field test are used in the model training, and the overall strategy is validated under PanoSim. Simulation results show that MTPM can accurately identify maneuvers such that the effective prediction on trajectories can be generated. TTC and threat index can be calculated timely. The proposed threat assessment strategy can not only assist collision avoidance systems to foresee dangerous situations but also eliminate false alarm to certain extent.

VI. Conclusion

The safety of self-driving vehicles has unique characteristics that differ from the safety of more established industries such as avionics, process control, and automotive. This book collection includes a sample of papers in the three types of safety that characterize self-driving vehicles: functional safety, SOTIF, and multi-agent safety. Traditional functional safety is still applicable for self-driving vehicles but is not sufficient. The safety that is associated with failures in the services provided by the perception system functionality, which are caused by performance degradation rather than random hardware or software failures, is referred to as SOTIF. Multi-agent safety involves addressing hazards due to actions of agents such as other vehicles, pedestrians, bicyclists, and in general any object that is external to the vehicle. Much of the literature on multi-agent safety is not addressed using a risk-based approach as it is the case with functional safety.

References

1. Thrun, S. et al., "Stanley: The Robot That Won the DARPA Grand Challenge," *Journal of Field Robotics* 23, no. 9 (2006): 661–692.

2. Urmson, C. et al., "Autonomous Driving in Urban Environments: Boss and the Urban Challenge," *Journal of Field Robotics* 25, no. 8 (2008): 425–466.

3. Cheng, H., *Autonomous Intelligent Vehicles: Theory, Algorithms, and Implementation* (London: Springer, 2011).

4. Wang, F.-Y. et al., "IVS 05: New Developments and Research Trends for Intelligent Vehicles," *IEEE Intelligent Systems* 20, no. 4 (2005): 10–14.

5. Cheng, H. et al., "Interactive Road Situation Analysis for Driver Assistance and Safety Warning Systems: Framework and Algorithms," *IEEE Transactions on Intelligent Transportation Systems* 8, no. 1 (2007): 157–166.

6. International Organization for Standardization, *Road Vehicles – Functional Safety*, ISO Standard 26262, 2011.

7. Koopman, P. and Wagner, M., "Autonomous Vehicle Safety: An Interdisciplinary Challenge," *IEEE Intelligent Transportation Systems Magazine* 9, no. 1 (2017): 90–96.

8. Avizienis, A. et al., "Basic Concepts and Taxonomy of Dependable and Secure Computing," *IEEE Transactions on Dependable and Secure Computing* 1, no. 1 (2004): 11–33.

9. Leveson, N.G., *Safeware: System Safety and Computers* (Addison-Wesley, 1995).

10. Wendorff, W., "Quantitative SOTIF Analysis for Highly Automated Driving Systems," *Conference Proceedings Safetronic 2017*, Stuttgart, Germany, 2017.

11. Thomas, J. et al., "An Integrated Approach to Requirements Development and Hazard Analysis," SAE Technical Paper 2015-01-0274, 2015, doi:10.4271/2015-01-0274.

12. Leveson, N.G., *Engineering a Safer World: Systems Thinking Applied to Safety* (MIT Press, 2012).

13. Young, W. and Leveson, N.G., "An Integrated Approach to Safety and Security Based on Systems Theory," *Communications of the Association for Computing Machinery (ACM)* 57, no. 2 (2014): 31–35.

14. Abdulkhaleq, A., Wagner, S., and Leveson, N., "A Comprehensive Safety Engineering Approach for Software-Intensive Systems Based on STPA," *3rd European STAMP Workshop, Conference Proceedings*, Amsterdam, The Netherlands, 2015.

15. Abdulkhaleq, A. et al., "A Systematic Approach Based on STPA for Developing a Dependable Architecture for Fully Automated Driving Vehicles," *4th European STAMP Workshop, Conference Proceedings*, Zurich, Switzerland, 2017.

16. Kopetz, *Real-Time Systems: Design Principles for Distributed Embedded Applications* (Kluwer Academic Publishers, 1997).

17. Koopman, P. and Wagner, M., "Toward a Framework for Highly Automated Vehicle Safety Validation," SAE Technical Paper 2018-01-1071, 2018, doi:10.4271/2018-01-1071.

18. International Electrotechnical Commission, "Functional Safety of Electrical/Electronic/Programmable Electronic Safety-Related Systems," IEC Standard 61508, 2010.

19. International Electrotechnical Commission, "Functional Safety – Safety Instrumented Systems for the Process Industry Sector," IEC Standard 61511, 2018.

20. Shalev-Shwartz, S., Shammah, S., and Shashua, A., "On a Formal Model of Safe and Scalable Self-Driving Cars," *Computing Research Repository (CoRR)*, arXiv:1708.06374 [cs.RO], 2017, [Online] http://arxiv.org/abs/1708.06374.

21. Kim, K. et al., "Design of Integrated Risk Management-Based Dynamic Driving Control of Automated Vehicles," *IEEE Intelligent Transportation Systems Magazine* 9, no. 1 (2017): 57–73.

22. Feth, P., Adler, R., Fukuda, T., Ishigooka, T., Otsuka, S., Schneider, D., Uecker, D., and Yoshimura, K., "Multi-Aspect Safety Engineering for Highly Automated Driving Looking Beyond Functional Safety and Established Standards and Methodologies," Gallina, B. et al. (Eds.). *SAFECOMP 2018, LNCS 11093*, 2018, 59–72.

The Development of Safety Cases for an Autonomous Vehicle: A Comparative Study on Different Methods

Junfeng Yang, Michael Ward, and Jahangir Akhtar
Birmingham City Univ

The Connected and Autonomous Vehicles (CAVs) promise huge economic, social and environmental benefits. The autonomous vehicles supposed to be safer than human drivers. However, the advanced systems and complex levels of automation could also bring accidents by tiny faults of hardware or errors of software. To achieve complete safety, a safety case providing guidance on the identification and classification of hazardous events, and the minimization of these risks needs to be developed throughout the entire development lifecycle process of CAVs. A comprehensible and valid safety case has to employ appropriate safety approaches complying with the automotive functional safety requirements in ISO 26262. The technical focus of present work is on the comparative study of different safety approaches, in particular, Failure Mode and Effects Analysis (FMEA) method and Goal Structuring Notation (GSN) method that have been employed to generate lists of hazardous events, safety goals and functional safety requirements at the vehicle level. A case study on the safety case development of INISIGHT autonomous vehicle has been carried out using the aforementioned methods. This case study covers the safety argument of battery and charging system that supply the whole electric power for INSIGHT vehicle. The safety of this systems has

CITATION: Yang, J., Ward, M., and Akhtar, J., "The Development of Safety Cases for an Autonomous Vehicle: A Comparative Study on Different Methods," SAE Technical Paper 2017-01-2010, 2017, doi:10.4271/2017-01-2010.

been assessed along with their potential for malfunction together with the layers of protection. The results and conclusions from case study analyses suggest the safety case of CAVs can be developed in a highly effective manner by employing a combined method of GSN and FMEA.

Introduction

Rapid growth in personal transport is frightening in terms of the spiraling number of injuries and deaths, global pollution and climate change. Back in 2009, 5.5 million accidents in the USA, involving 9.5 million vehicles, killed ~34k people and injured >2.2M others, including 240k hospital admissions [1]. In addition, cars and trucks are estimated to cause 20% all U.S. CO_2 emissions [2]. For the exploding numbers of cars in the developing world, the statistics are even more terrifying. The CAVs equipped with more sensors to detect other road uses and pedestrians, and much higher levels of computer control promise huge reductions in accidents, congestion, and pollution. For example Google claim their driverless car could reduce accidents by 90%, wasted time and fuel by 90%, and massively increase the utilization of cars, meaning fewer cars overall [3]. With a huge market worth in view, every major car manufacturer in the world is developing CAVs. One estimate for sales of autonomous vehicles is 95 million per year by 2035 [4]. The IEEE predicts that autonomous vehicles could be as much as 75% of the market by 2040 [5]. The automotive industry makes a substantial contribution (>£60Bn) to the UK economy, and is expected to see considerable growth in the next decade and more.

One of the earliest reports on autonomous vehicle appeared on 1948 which concerned the development of cruise control in vehicles [6, 7]. Since then the work has been developed by many researchers to include areas such as mechanical antilock braking, electrical stability control, laser based cruise control, pre-crash mitigation. Some of the first autonomous car projects in the 1980s were the Navlab (1980) and the ALV (Autonomous Land Vehicle) in 1984 that were organized by Carnegie Mellon University (CMU). They have continued to develop the autonomous car since then. Recently the CAVs industry keeps blooming and many companies including Mercedes-Benz, General Motors, Continental Automotive Systems, Autoliv Inc., Bosch, Nissan, Toyota, Audi, Volvo, Google and Tesla have developed autonomous cars [8, 9].

Figures for UK's CAVs development show a similar trends. Within UK, the Centre for Connected & Autonomous Vehicles, CCAV, has been established to help ensure UK's world leadership in developing and testing connected and autonomous vehicles. Since 2015, CCAV has continuously funded a series of projects, e.g. GATEway, Venturer, UK Autodrive, INSIGHT, i-MOTORS and FLOURISH, on intelligent mobility research and development. Among them, the INSIGHT project aiming to develop driverless shuttles with a particular focus on improving urban accessibility for disabled and visually-impaired people will be thoroughly introduced. And the safety case developed for the INSIGHT pod will be discussed as the case study in the follow in g sections.

The INSIGHT [10] project is a collaborative project to develop existing autonomous vehicles for safe, slow speed (max 15 MPH) operation in pedestrian areas and pavements, with connectivity not only to control and manage the vehicles, but also for innovative data collection and presentation applications that interact with users and other customers of the systems. An existing electric connected & autonomous vehicle design [11] has been upgraded with advanced sensors to detect and recognize pedestrians, cyclists,

FIGURE 1 General exterior views of the INSIGHT vehicle.

FIGURE 2 General interior views of the INSIGHT vehicle.

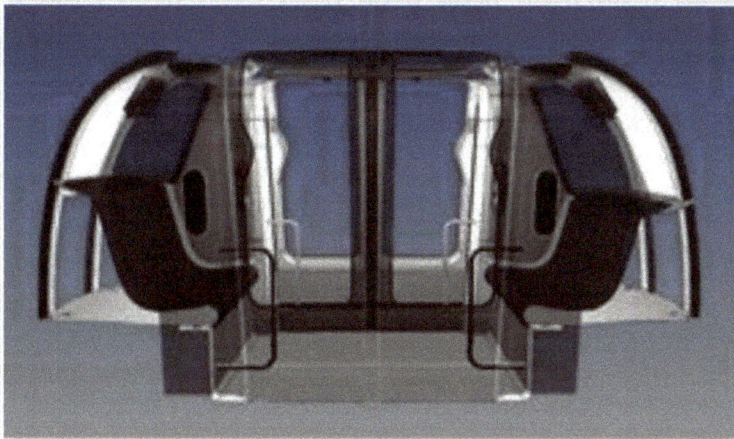

mobility scooters, and other vehicles on adjacent roads. These detection capabilities will enable more advanced decision making and a more nuanced approach to way finding, and a smoother ride rather than the simple start/stop common in such systems today. The general view of INSIGHT autonomous vehicle can be seen in Figure 1.

The INSIGHT pod vehicle is driverless and self-steering (autonomous driving SAE Level 5 [12]), electrically-powered light-weight vehicle designed to carry up to four people and their luggage (including pushchairs and bulky items), see Figure 2 for the general interior view. While the INSIGHT pod is suitable for almost any age group, it has been designed with inclusive at its heart. The vehicle has wheelchair access and, INSIGHT will look specifically at its use by the elderly and those who need assistance in transport, for example the visually impairment. The pod will not just assess the physical passenger experience, such as internal comfort and safety, but also the supply journey information such as calling response times, destination, connections and other support information, all delivered by a human voice interface.

The project activities require a safety case to be made before commencing in order to provide assurance that any reasonable residual risks have been minimized and where possible avoided all together. In addition, a safety case based upon the road vehicle functional safety standards is also required to demonstrate that the vehicle can be safely and reliably driven. The typical approaches documenting safety cases include textual format, tabular form e.g. using FEMA, graphical notations, e.g. GSN. All these approaches have been employed to formulate the safety cases for various autonomous vehicles. For instance, LUTZ pathfinder automated vehicle [13] has developed a defendable safety case using FEMA approach together with a tailored application of ISO26262 automotive functional safety standard [14], and concluded the use of human intervention is required for trials. Another similar project, ULTra CAVs providing personal rapid transition between the T5 Business Car Park and Terminal 5 of Heathrow Airport generated a safety case using a combined method of FEMA and GNS. The ULTra CAVs has been safely running since 2011 and delivered in excess of 3.5million passengers, which provide a strong and convincing evidence on that successful safety case.

The INSIGHT vehicle is based upon an existing design ULTra CAVs, where the vehicle supposed to operate in an unconstrained pedestrian area instead of is confined to a well-defined purpose built track. It is necessary therefore that an appropriate safety is developed for INSIGHT vehicle. The present work seeks to explore various approaches for developing a safety case for INSIGHT vehicles. These approaches will incorporate with the general safety management follows a diverse set of legislation and guidance, e.g. SAE J3018 and J3061 [15], UK Code of Practice for Testing Driverless Car [16] and the Road Traffic Act [17]. The performance of various approach will be compared and analyzed.

Vehicle Layout and ISO26262

Vehicle Control System and Propulsion System

In order to allow level 5 autonomous vehicle operation, an autonomous control system has been developed for INSIGHT pod vehicle. This autonomous control system consists of situational awareness system, central control system facilitating dynamic path planning and decision making, and vehicle movement control system. The smart sensor module ((a group of sensors, e.g. long- and short-range radars, front-, rear-, and side-stereo-cameras and ultrasonic sensors)) connected via wired Ethernet to autonomous vehicle central control system to improve path planning and decision making. The module integrates steering, brakes and E-Motor which respond to demand from the vehicle control system, transmitted via a CAN bus. Figure 3 presents the basic control system architectural of an INSIGHT vehicle. Note that this basic control system is expressed exclusive of the environmental monitoring sensor system, human-machine-interaction control and 4D tactile system for a clearer view.

The INSIGHT vehicle has two Li-ion battery units, high voltage (48V) electric power for traction power system, and low voltage (24V) electric power for vehicle control system and door actuation system. The typical propulsion system for an INSIGHT vehicle is shown in Figure 4. As can be seen, the motor control module converts the 48V DC battery power into low voltage 3 phase AC power while simultaneously controlling motor torque speed and direction. The AC E-Motor drives the vehicle through the front wheels via a fixed ratio transmission and a differential mounted in a transaxle. Note that, the nominal system voltage is restricted to 48 volts to minimize the shock risk.

FIGURE 3 Basic functional topology of an INSIGHT vehicle control system.

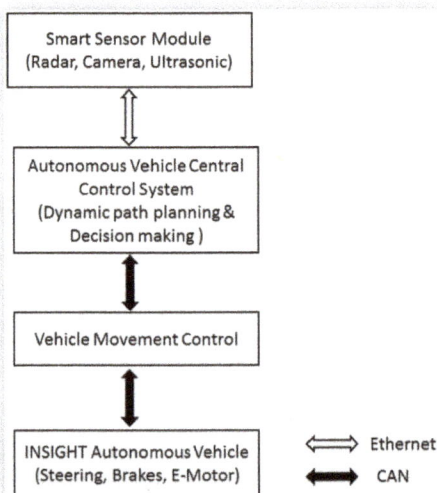

FIGURE 4 Typical propulsion system schematic for an INSIGHT vehicle.

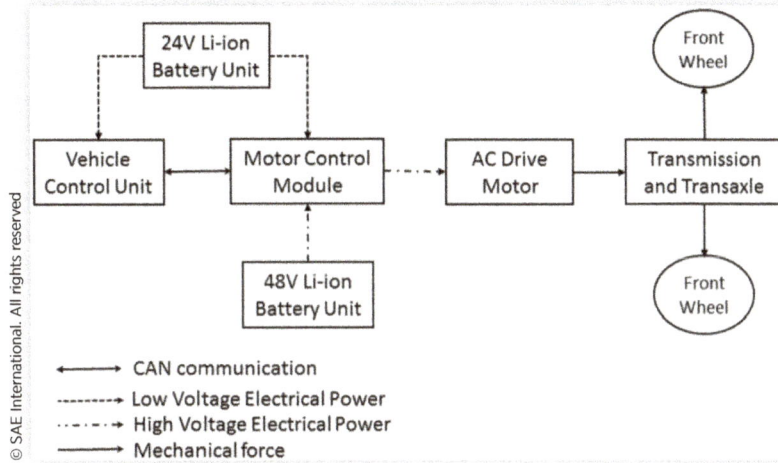

ISO26262 Road Vehicle Functional Safety Standard

ISO26262 is an adaption of IEC 61508 [18] to meet specific needs of automotive industry. It is the first comprehensive standard that addresses safety related automotive systems comprised of electrical, electronic, and software elements that provide safety related functions. It seeks to address the following important challenges in today's road vehicle technologies: the safety of new electrical and electronics hardware and software functionality in vehicles; the trend of increasing complexity, software content, and mechatronics implementation; the risk from both systematic failure and random hardware failure. It also provides guidance on how to avoid risk in creating safety-critical systems and regulates critical testing processes.

ISO 26262 defines a safety case as an "argument that the safety requirements for an item are complete and satisfied by evidence compiled from work products of the safety activities. Figure 5 present a system lifecycle approach (V-model) used throughout a safety case. This lifecycle model represents the development of the system from first concept to operation. The concept phase (Part 3) refers to the initial big picture of

FIGURE 5 System lifecycle approach to safety: used throughout a safety case.

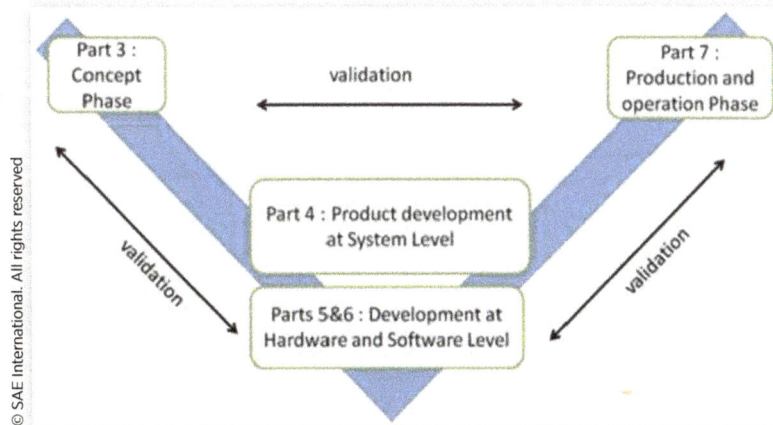

autonomous vehicle in terms of styling and functionality, etc. Parts 4-6 refers to the vehicle develop and software/ hardware development. Part 7 refers to the final product. Validations refer to various trials, e.g. commissioning test (vehicle shakedown), on-road tests of hardware/software, and trial under a public pedestrian area at different phases. The V-shape is due to the fact that the testing and verification steps are performed in reverse order from design and implementation.

Failure Model and Effects Analysis Method

Failure Mode and Effects Analysis (FEMA) method developed initially for analyzing malfunctions of military systems [19] uses a structured, systematic spreadsheet to documents all the possible failures, risk assessment and management strategies in a design, a manufacturing or assembly process, or a product or service.

The typical FMEA spreadsheet captures all systems/components information including Items, Functional Requirements, Failure Modes, and Causes of Failure. Each possible Cause of Failure has an associated Risk to it which is derived from its Occurrence & Severity. After this the first focus is on Design actions, which means going through the Causes of Failure and mitigate these risks down to the lowest possible level. After the design actions have been considered, the part should be ready for validation through either physical testing or rationale, proving its robustness and ability to meet the safety performance requirements. During Validation, design review, customer reviews/testing issues/concerns may be raised. These issues or failures need to be fed into FMEA ensuring that additional risk is added to the parts of concern proving that we have mitigated that failure and are fit to continue the validation process. The example spreadsheet of FMEA method was given below

Taking the advantages of intuitively clear and high viability, FMEA method has been further developed and adopted by the aerospace and automotive industries.

Goal Structuring Notation Method

Goal Structuring Notation (GSN) method [20] is a graphical notations for the representation of arguments, which was first proposed by T. P. Kelly [21] under the inspiration of Toulmin's argument model [22]. GSN method employs a simple notation of argumentation structures that have been proven to be effective for provides objective safety evidence, therefore has been widely used for developing safety cases for the industrial use and research purpose. Recently, GSN method has been incorporated into ISO26262 to satisfy the critical safety assurance of automotive systems, e.g. start/stop system, and EPS system [23, 24, 25, 26, 27, 28].

Typically, GSN method consists of a group of symbol of notations linked by directional arrows explicitly representing the individual elements (safety goals, solution, context and strategies) of an argument and the relationships between these elements, such as rectangular boxes for safety target, ovals for assumption or justification, circle for evidence (solution), parallelogram-shaped boxes for strategy (argument),

TABLE 1 Example risk assessment represented using FMEA spreadsheet

Item	Function Requirement	Failure Model	Casual Factor	Immediate Consequences	Severity	Frequency	Exposure	Hazard Ranking	Mitigation
Battery	supply electric power for E-motor	supply insufficient electric power	over- heating; degradtion	vehicle lose power	3	2	2	6	BMS monitory battery voltage and temperature

FIGURE 6 An example safety argument represented using GSN.

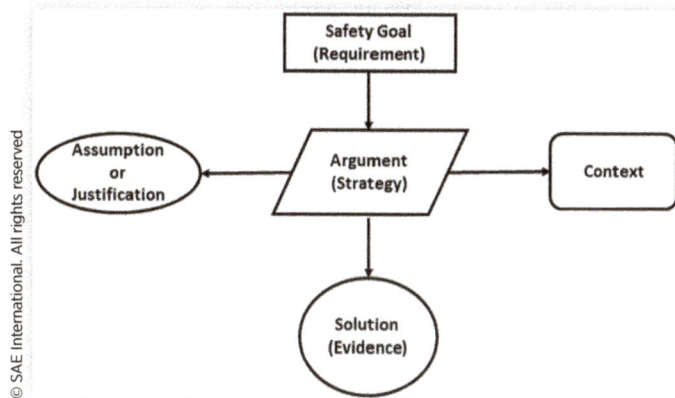

rounded-end boxes for context (additional information). An example safety argument constructed using GSN is given below.

As shown in Figure 6, the safety goal of targeted system needs to be claimed by identifying the possible hazards and mitigating them through sufficient and appropriate evidences. Due to the complexity of system, the top-level safety goal usually has to be decomposed into sub-level goals and this decomposition may continue until sub-level claim and evidences asserted.

Both FMEA and GSN have been widely used for the risk assessment of automotive industries, and proven to be valid and capable methods. The following sections provide a detailed description on a safety case for INISHGT autonomous vehicle constructed using FMEA and GSN methods incorporating with ISO26262 road vehicle functional safety standard. In addition, the performance appraisal of these method provides guidance toward a valid and defendable safety case.

Safety Case Development

In the present work, the safety case investigates and documents all the hazards and risks associated with autonomous vehicle, including mechanical system, electric hardware, application and embedded software, communications, health and safety of passengers, roads users, pedestrians, risk assessments, safety systems of work and insurance and liability.

A valid safety case for an autonomous vehicle consists of four main inter-dependent components, namely:

- Safety target that must be addressed to assure vehicle safety.
- Evidence for the safety target obtained from study, analysis and test of the vehicle system.
- Argument showing how the rationale indicates compliance with the safety target.
- Context identifying the basis for the argument presented.

A set of safety targets for the vehicle commissioning tests is generated with the objective of achieving acceptable safety considering the prototype nature of INSIGHT vehicle. Based on the goals of the safety work, the principles of safety process rationale argument were chosen as:

1. Hazard Generation. Identifying the vehicle operational situation and the possible hazardous events associated with safety targets.

TABLE 2 ASIL Determination per ISO26262:2011

ASIL Classification		C1	C2	C3
S1	E1	QM	QM	QM
	E2	QM	QM	QM
	E3	QM	QM	A
	E4	QM	A	B
S2	E1	QM	QM	QM
	E2	QM	QM	A
	E3	QM	A	B
	E4	A	B	C
S3	E1	QM	QM	A
	E2	QM	A	B
	E3	A	B	C
	E4	B	C	D

2. Risk Assessment. Classifying each hazard in terms of frequency of occurrence, severity of resulting harm and controllability of hazard, and determining the Automotive Safety Integrity Level (ASIL) of system by considering the SAE J2980 standard [29].

3. Hazard Management. Addressing safety requirements through an appropriate combination of system design in accordance with the ASIL indicated.

An ASIL shall be determined for each hazardous event based on its severity level (S1-S3), probability of exposure (E1-E4) and controllability level (C1-C3) in accordance with Table 2. The number 1 represents lowest level and 4 the highest one. The classification of severity, exposure and controllability are given in SAE J2980.

As can be seen, Four ASILs are defined: ASIL A, ASIL B, ASIL C and ASIL D, where A representing the least stringent level and D the most stringent level. QM indicates quality management system can be sufficient to develop element(s) that implement the safety requirement allocated to the intended functionality. Or it can support the rationale for the independence between the intended functionality and the safety mechanism.

Figure 7 summarizes the safety process rational argument employed in the present work. If a system has high ASIL and subjects to the constraints, its safety requirement can be decomposed by multiple redundant subsystem working together, each with a lower ASIL. This process is so-called ASIL decomposition that allows the best safety strategies to be developed efficiently.

Case Study

The INSIGHT autonomous vehicle has two Li-ion battery units mounted on to the battery tray of the rear. The battery and charging system supply as one of the most important systems supply the whole electric power for the INISIGHT vehicle. These batteries incorporates an on-board Battery Management System (BMS) in the vehicle. The BMS has the feature of measuring cell voltage and temperature, performing cell balancing function and monitoring the cell fault conditions, providing these information to the external systems via CAN. Since the Lithium-ion batteries contain flammable electrolyte and may pose a fire/explosion and other hazards when it's overheated or short-circuited. Thereafter the safety case must include the risk assessment on battery and charging system. This section takes the battery and charging system as the case study to examine the product-based safety rational argument.

Table 3 present the FMEA-based safety argument for the battery and charging system. The potential malfunctioning behavior relevant to the battery and charging

FIGURE 7 Safety process rationale argument.

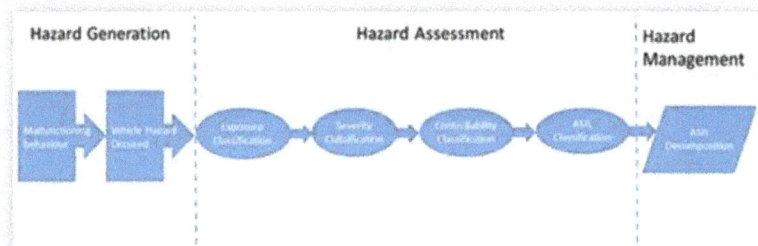

TABLE 3 FMEA-based safety argument for the battery and charging system of INSIGHT vehicle

Entry id	Item	Cause of failure	Consequence	Mitigation	Controllability	Severity	Frequency	ASIL Ranking	General Comments
1	Battery/charger	Electric current leakage	Electric shock	Nominal system voltage is restricted to a non-lethal level (48volts). No passenger exposure to high voltage. No circuits used within the passenger compartment operate at voltages above 24V. Vehicle charging contacts are mounted under the vehicle and are inaccessible to passengers. Neither vehicle nor the mounted charging contacts are live when they are not connected together	C1	S3	E1	QM	
2	Battery/charger	Resistive connection	Overheating	Temperature sensing employed to detect excessive temperature at charging contacts. Purpose designed connectors used for all high current connections.	C1	S1	E3	QM	
3	Battery/charger	Rapid charging or discharging	Fire/explosion	Battery protection fuse mounted within battery pack. The battery pack is physically separated from the passenger compartment by bulkheads.	C1	S3	E3	A	
4	Battery/charger	Battery heating	Fire/explosion	Battery chargers automatically control charging to avoid overcharging. Vehicle controller continuously monitors individual battery voltage and temperature; it will stop charging if overcharging or over temperature fault conditions are detected.	C1	S3	E2	QM	

FIGURE 8   GSN-based Safety goal rational argument for a battery /charging system of INSIGHT vehicle.

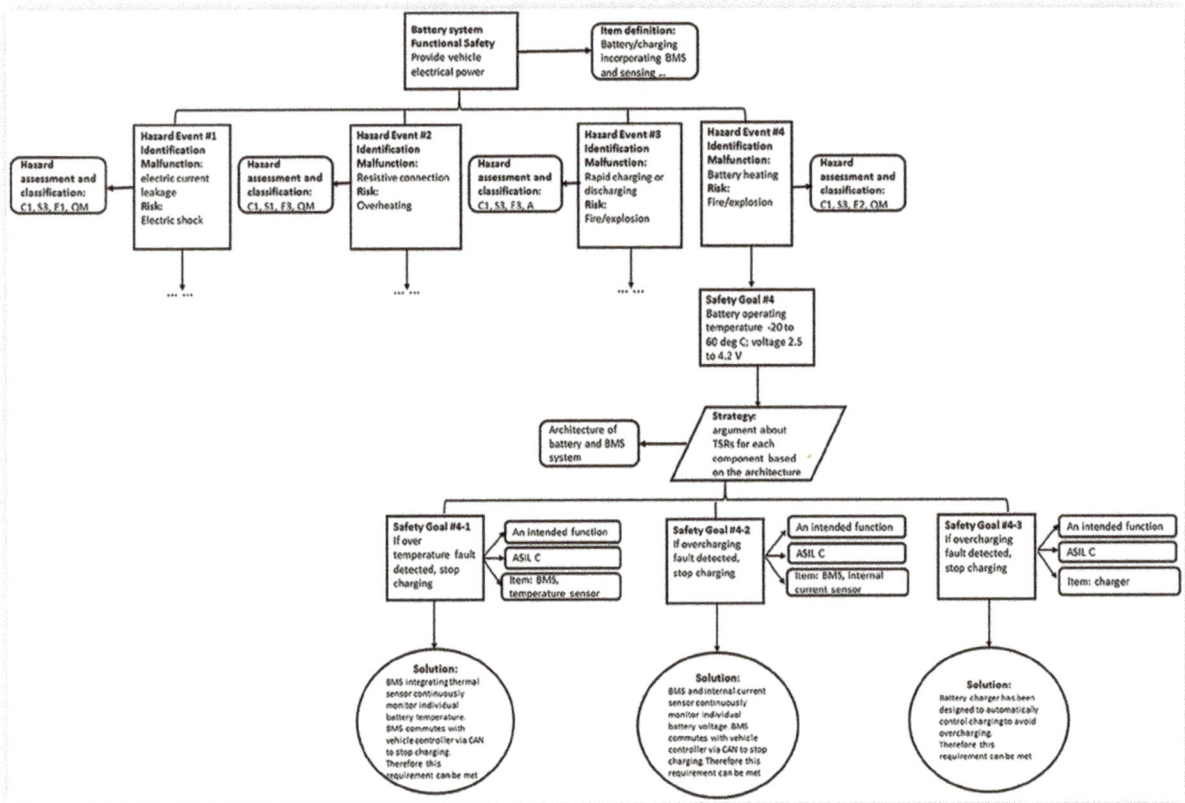

system and potential hazards have been identified. Then severity level, controllability and probability of exposure of this vehicle hazard and ASIL level are classified for each hazards. Failure of the battery and charging system is clearly undesirable. However system design features mitigate the risk associated with such failures to acceptable levels. Figure 8 shows GSN-based product argument structure for the battery/charging Systems Safety argument. This valid argument (strategy) is supported by evidence in compliance with ISO 26262.

As can be seen clearly, FMEA is better for documenting the evidence and context. But it seems hardly to convey all necessary information effectively along. While, GSN are better for presenting a decision making process, especially the decomposition of safety goal on a complex system which involves a number of risks. However, for a complex system, the GSN has to cover a number of sub-goals which may cause an intricate GSN structure and make it difficult to follow.

Conclusions

Safety has been a prime consideration throughout the development of the INSIGHT autonomous vehicle. The present work has developed a safety case in order to provide assurance that any reasonable residual risks have been avoided during the vehicle commissioning test. This safety case builds in accordance with the ISO26262:2011 road vehicle functional safety standard. The principle conclusion of safety assessment is that no features of the INSIGHT design concept and operating concept would indicate that the level of safety of the INSIGHT system would be unacceptable. The INSIGHT vehicle is therefore acceptably safe to commence operations.

A diverse approach to the assessment of the safety of the system has been adopted, including assessment against FMEA, GSN and SAE guidance, and quantified risk assessment. The risks associated with all the identified hazards are considered to be in the tolerable or acceptable risk categories. A case study on a battery and charging system of INSIGHT vehicle demonstrates the typical structure of safety goal rational argument. The analysis results indicated that the approach adopted was appropriate.

Regarding the performance appraisal of FMEA and GSN methods, it found FMEA is better for documenting the evidence and context. But it seems hardly to convey all necessary information effectively along. While, GSN are better for presenting a decision making process, especially the decomposition of safety goal on a complex system which involves a number of risks. Hence, a combined method of FMEA and GSN is suggested to contracture a valid and defendable safety case in an efficient and effective manner.

In terms of future work, we would like to continue our safety case to explore the aforementioned method to address the unique aspects of CAVs, e.g. navigation system and decision-making system, which have not been thoroughly discussed.

Contact Information

Junfeng Yang, PhD
School of Engineering and the Built Environment
Faculty of Computing, Engineering and the Built Environment
Birmingham City University
City Centre Campus
Millennium Point
Birmingham B4 7XG
United Kingdom
Phone: +44 (0)121 300 4293
Junfeng.Yang@bcu.ac.uk

Acknowledgments

This project was funded by Innovate UK (grant agreement No. 102583) and supported by the Centre for Connected and Autonomous Vehicles, UK.

Definitions/Abbreviations

CAV - Connected and Autonomous Vehicle
FMEA - Failure Mode and Effects Analysis
GSN - Goal Structuring Notation
ASIL - Automotive Safety Integrity Level
BMS - Battery Management system

References

1. http://www-nrd.nhtsa.dot.gov/Pubs/811363.pdf.

2. http://www.ucsusa.org/our-work/clean-vehicles/car-emissions-and-global-warming.

3. http://www.google.co.uk/about/careers/lifeatgoogle/self---driving-car-test-steve-mahan.html.

4. http://www.navigantresearch.com/research/autonomous---vehicles.

5. http://www.ieee.org/about/news/2012/5september_2_2012.html.

6. "keesler news," keesler, 1 1 1948, [Online] accessed October 13, 2016, http://www.keesler.af.mil/AboutUs/FactSheets/Display/tabid/1009/Article/360538/history-of-keesler-air-force-base.aspx.

7. Ioannou, P.A. and Chien, C.C., "Autonomous Intelligent Cruise Control," *IEEE Transaction on Vehicle Technology* 42, no. 4 (1993): 657-672.

8. Thorpe, C., Hebert, M., Kanade, T., and Shafer, S., "Toward Autonomous Driving: The CMU Navlab," *IEEE* (1991): 31-41.

9. Thorpe, C., Hebert, M., Kanade, T., and Shafer, S., "Vision and Navigation for Carnegie-Mellon Navlab," *IEEE* 10, no. 3 (1988): 362-373.

10. Centre for Connected and Autonomous Vehicles, UK, *INSIGHT* Project, accessed January 2016, http://insightcav.com.

11. Heathrow Pod, accessed July 2015, http://www.ultraglobalprt.com/wheres-it-used/heathrow-t5/.

12. SAE International Surface Vehicle Information Report, "Guidelines for Safe On-Road Testing of SAE Level 3, 4 and 5 Prototype Automated Driving Systems (ADS)," SAE Standard J3018, Iss, Mar. 2015.

13. Peters, A., "Safety of the LUTZ Pathfinder Automated Vehicle," *22nd ITS World Congress, Paper number ITS-2427*, Bordeaux, France, October 5-9, 2015.

14. *ISO 26262 -Road Vehicles -Functional Safety. Parts 1 to 12.*

15. SAE International Surface Vehicle Information Report, "Guidelines for Safe On-Road Testing of SAE Level 3, 4 and 5 Prototype Automated Driving Systems (ADS)," SAE Standard J3018, Iss, Mar. 2015.

16. UK Department for Transport, "The Pathway for Driverless Car: A Code of Practice for Testing," 2015.

17. UK Government, "Road Traffic Act," 1991.

18. IEC 61508, "Functional Safety of Electrical/Electronic/Programmable Electronic Safetyrelated Systems."

19. United States Department of Defense, "MIL-P-1629 - Procedures for Performing a Failure Mode Effect and Critical Analysis," Department of Defense (US), MIL-P-1629, November 9, 1949.

20. Goal Structuring Notation Working *Group*: GSN Community Standard Version 1. http://www.goalstructuringnotation.info/, 2011.

21. Toulmin Stephen, E., *The Uses of Argument* (Cambridge University Press, 1958).

22. Kelly, T.P., "Arguing Safety - A Systematic Approach to Safety Case Management," DPhil Thesis YCST99-05, Department of Computer Science, University of York, UK, 1998.

23. Kelly, T. and Weaver, R., "The Goal Structuring Notation-A Safety Argument Notation," *Proceedings of the Dependable Systems and Networks 2004 Workshop on Assurance Cases*, July 2004.

24. SAE International Surface Vehicle Recommended Practice, "Cybersecurity Guidebook for Cyber-Physical Vehicle Systems," SAE Standard J3061, Iss, Jan. 2016.

25. Palin, R. and Habli, I., "Assurance of Automotive Safety: A Safety Case Approach," *SAFECOMP 2010*, Vienna, Austria, 2010.

26. Palin, B., Ward, D., Habli, I., and Rivett, R., "ISO 26262 Safety Cases: Compliance and Assurance," *IET Intl. System Safety Conf.* (2011).

27. Habli, I. et al., "Safety Cases and Their Role in ISO 26262 Functional Safety Assessment," *32nd International Conference on Computer Safety, Reliability, and Security*, Toulouse, France, 2013.

28. Matsuno, Y., "D-Case Ediotor," http://www.il.is.s.u-tokyo.ac.jp/deos/dcase/.

29. SAE International Surface Vehicle Recommended Practice, "Considerations for ISO 26262 ASIL Hazard Classification," SAE Standard J2980, May 2015.

A Means of Assessing the Entire Functional Safety Hazard Space

Daniel Aceituna
DISTek Integration Inc

The goal behind Functional Safety is to anticipate the potential hazard scenarios (a.k.a. harm sequences) that a system may produce and address those scenarios in such a way as to mitigate or even eliminate them. A major challenge in determining hazard scenarios is trying to assess an adequate amount of scenarios, considering the large size of a hazard space. Typically assessing the entire hazard space is difficult to achieve, resulting in the possibility of overlooking some critical scenarios that can result in harm to either system operators, system by-standers, or both. In this paper we will explore a rule-based approach for concisely describing hazard scenarios, which could potentially enable us to examine the entire hazard space in a short amount of time. Our approach, called Hazard Space Analysis, combines three key activates: determining hazard scenarios, assigning a risk factor to those scenarios, and mapping those hazard scenarios directly to safety rules. We will detail the approach, show how the approach could be automated, and present a simple aviation related example that demonstrates the approach's potential in enabling stakeholders to explore a large hazard space.

CITATION: Aceituna, D., "A Means of Assessing the Entire Functional Safety Hazard Space," SAE Technical Paper 2017-01-2056, 2017, doi:10.4271/2017-01-2056.

Introduction

Functional Safety (FS) has been defined in various ways, such as, a system's automatic ensuring of a safe, non-hazard state, or part of the overall safety that depends on a system or equipment operating correctly in response to its inputs [21]. For our purposes, we will view Functional safety as the practice of assessing a system's potential hazards and reducing those hazards to an acceptable risk. Various ISO documents define how the practice of Functional safety should be performed in the auto industry (ISO 26262), Agricultural industry (ISO 25119), Machinery (ISO 13849), Aerospace (ARP 4761), and the electronics industry in general (IEC 61508), to name five. Whichever the industry, there are some common FS practices that are applied to the system being safe-guarded, in particular, the practice of assessing potential hazardous situations (or harm sequences), which we will refer to as Hazard Scenarios (HS). These hazard scenarios are used to determine the safety functions needed to mitigate the hazards. Safety functions can be implemented by either a separate protection system, or integrating the safety functions into the system being safe guarded, or a combination of both. The assessment of Hazard Scenarios is the focus of this paper. In particular, we present a framework for potentially addressing the problem of adequately assessing a system's hazard space. We define a hazard space as the complete set of hazard scenarios that a system can potentially experience. Determining a system's entire hazard space is difficult due to the large combination of things that can go wrong.

In typical FS practice, engineers and stakeholders must assess the various hazard scenarios that can be produced or experienced by a system. This involves assessing several artifacts, such as, the system in various operational situations, potential environment states the system will operate in, and the various nominal and off nominal interactions with human operators and/or human bystanders in proximity to the system (just to name a few). Compiling the stated artifacts present their own challenges, however, a bigger challenge is conceiving of those artifacts in all possible combinations. We view the standard FS practice of compiling hazard scenarios as additive; the hazard space is built by assessing and adding one scenario at a time. Arguably, the potential to overlook a key relevant hazard scenario is high when using an additive approach. Secondly, if we assume that the total number of relevant scenarios is closer to the upper end of the hazard space size, then it may be feasible to some degree to start with nearly the entire space and trim down to the relevant number of scenarios, than to start from zero scenarios and work up to the relevant number. We view the act of trimming down as subtractive. On the assumption that a subtractive approach could result in more coverage than an additive approach we explore a means of assessing as much of the hazard space as possible using a subtractive approach we call Hazard Space Analysis (HSA).

HSA utilizes a rule-based notation that describes a hazard scenario as a transition from a system state (operational situation) to a hazard state as caused by one of several possible action/states. The rule-based format also allows for a concise, high level, description of a hazard scenario that maps directly into a safety requirement. Each rule consist of three artifacts: Operational Situation (OS), Hazard State (HS), and a hazard cause which in turn can be classified as either an Environmental State (ES), an operator Nominal Behavior (NB), an operator Off-Nominal Behavior (ONB), or a Component Failure (CF). These various classifications allow for the partitioning of the hazard space, as well as a framework for the potential combining of hazard causes. In HSA the hazard causes ES, NB, ONB, and CF are assess by engineers and stakeholders, whereas the hazard scenarios are automatically generated from the assessed causes. The computer generation of scenarios makes it possible to produce a large number of hazard scenarios that make up the system's hazard space. HSA also allows for the risk assessment of the scenarios based on hazard severity and probability of occurrence of each scenario. In summary, HSA provides a means to potentially examine the entire hazard space, reduce that space to a relevant, risk assessed, subset of hazard scenarios, and then use those scenarios to create a set of safety rules that raise the necessary questions to develop the appropriate safety functions during the design phase of a protection system. In the following sections we will provide background pertaining to our approach, and its related work. Other sections will also detail our approach and apply it to an aviation centered example; a discussion then follows.

Background

The desire for safe operating environments can be traced back to the 1930's when the United States recognized the need to establish job-related safety laws. The need for safety increased into the 1970's with the enactment of the Occupational Safety and Health Act. As automation and embedded systems became more prominent, the potential for hazardous situations that could harm a human being increased as well, calling for even more laws and practices surrounding safety. Control Systems (such as ECUs) have become more safety critical, as they integrate with more and more EUCs (Equipment Under Control). Society is presently surrounded by systems can potentially produce hazardous situations; with the potential increasing with the advent of autonomous vehicles and AI directed robotic systems. This has prompted the need for a systematic and standardized means of designing systems that exhibit safe operation; thus the need for practicing Functional Safety.

There are various ISO and IEC documents that define how the practice of Functional Safety should be performed in industries such as the auto industry (ISO 26262), Agricultural industry (ISO 25119), and Machinery (ISO 13849), to name three. Most modern industries rely on electronic control systems and related software. This requires the functional safety of Electrical/Electronic/Programmable Electronic Safety related Systems (E/E/PE or E/E/PES), as specified in the IEC 61508 standard. IEC 61508 is considered a generic standard that satisfies the needs of all industry sectors; it can also serve as the basis for drafting new standards to a particular industry. Section 3.1.12 of IEC 61508 defines Functional Safety as *"part of the overall safety relating to the equipment under control and the control system that depends on the correct functioning of the electrical, electronic and programmable electronic safety-related systems and other risk reduction measures".*

In a more concise and practical definition, Functional Safety (FS) can be defined as a system's automatic ensuring of a safe, non-hazard state. According to IEC 61508, there are 16 safety life cycle phases that should be followed when applying Functional Safety to a system consisting of an ECU controlling an EUC. The IEC 61508 phases translate into the following seven steps:

1. Defining the System (being designed and safe-guarded)
2. Determine the System's Safety Implications
3. Hazard and Risk Analysis (based on severity and probability)
4. Eliciting Safety Requirements (to reduce the assessed risk)
5. Designing Safety Functions (from the safety requirements)
6. Implementing the Safety Functions (internal or external to the system being safe-guarded)
7. Validation of Safety Functions

The IEC 61508 standard further breaks down some of the seven steps into sub-steps, but for the sake of this article we have generalized the steps. The IEC 61508 also addresses the concept of Safety Integrity and Safety Integrity Levels (SIL). However, we will forgo discussions of SILs since our focus is Hazard Scenarios (steps 2, 3 and 4), and SILs are beyond the scope of this paper.

The seven steps describe a process that parallels the development of the system being protected against hazards. This counters the assumption that designing a functionally reliable system automatically makes the system safe. Instead of assuming that a well-designed system is safe, Functional Safety takes a risk-based approach, in which safety activities are based on first understanding the hazard risks posed by the system being protected. The overall idea is to assess the hazard implications of a design, assign a risk factor to those hazards and reduce those risks via the system's design, or by designing an external system that acts as a safety monitor or protection system, as illustrated in Figure 1. This paper focuses on steps 2, 3, and 4, with emphasis on addressing the entire hazard space. Referring back to the seven steps derived from IEC 6108, we next describe steps 2, 3 and 4 in greater detail.

Step 2 focuses on assessing how the system-to-be (Figure 1, [A]) being protected can produce a hazardous situation harmful to a human being (Figure 1 [B]). This is typically achieved by conceiving of the different failure modes that can occur and how those failures can produce hazardous situations. The hazardous situations can be described as harm sequences (referred to hazard Scenarios in this paper), which detail to various degrees, how a component failure, human error, or environmental condition can result in harm. The Hazard Scenarios are conceived one at a time, as they occur to the engineer/stakeholders, resulting in an additive approach to creating a list of scenarios. This presents a challenge when trying to examine the majority, if not all, of the hazard space.

Step 3 involves assessing the risk associated with each hazard scenarios and assigning a risk value. The risk value is derived by determining the severity of the hazard scenario and the probability of that scenario occurring.

In Step 4 the scenarios with the unacceptable risk values are translated into a safety requirement (Figure 1, [C]) used to specify which hazards need to be mitigated to an acceptable risk value.

FIGURE 1   Representation of IEC 61508 process.

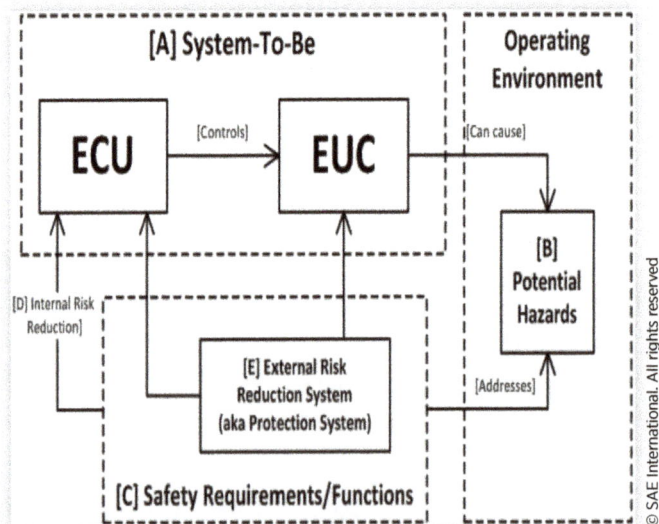

The three steps described are typically conducted manually, which can discourage many from examining a large number of scenarios. This discouraging effect became part of our motivation behind partly automating steps 2 and 3. Another motivating reason was the realization that an additive approach is less likely to achieve an examination of the complete hazard space. No matter how many scenarios one adds, there is always the possibility of there being one more scenario that is overlooked; an additive approach would be open-ended. Thus, we have the motivation for applying a subtractive approach, in which we computationally create the entire hazard space and then eliminate those scenarios that are not relevant and/or have an extremely low risk value. A subtractive approach is inherently not open-ended.

There is also the argument that the number of relevant hazard scenarios is more likely to be closer to the upper limit of the hazard space size. Therefore it makes more sense to start from that upper limit and reduce, rather than to start from zero and increase; as in the additive case. One potential drawback with a subtractive approach is the very large number of scenarios that an engineer may end up examining as he/she reduces the hazard space to a relevant size. However, we have added provisions, via a partitioning scheme, for reducing the cognitive workload on the engineer. We also feel the tradeoff in examining the entire space is worth the extra time, especially when that extra time can be minimized. Since it arguably makes sense to use a subtractive reduction of the hazard space, we now consider whether prior research has been conducted related to a system's hazard space and its reduction.

Related Work

An informal survey of SAE papers, as recent as 2017, revealed fifty-four technical and journal papers related to the topic of Functional Safety. Of the fifty-four papers, twelve focused on safety requirements, *e.g.* [5], ten addressed the ISO 26262 standard, *e.g.* [6], three addressed electromechanical braking systems, *e.g.* [7], six concerned electric vehicles, *e.g.* [8], three looked into using a model-based approach to Functional Safety, *e.g.* [9], two focused on functional safety for microcontrollers, *e.g.* [10], and finally fourteen dealt with other miscellaneous aspects of functional safety. There were no papers found that specifically addressed the hazard space and its inherent challenges.

Other repositories of technical and journal papers such as CiteSeerX, Microsoft Academic Search, Google Scholar, and IEEE Xplore, revealed two thousand plus papers related to Functional Safety. These represented various research directions in Functional Safety. Some proposed the use of formal methods in functional safety analysis [1, 2, 3]. Others expanded on prior techniques, as in the case of adapting hazard identification techniques, such as HAZOP, to safety requirements scenarios [4], the semi-automatic assembling of aviation safety cases [15, 19], the bridging of the gaps between model-based engineering safety driven design [16], and the managing of product variability in Functional Safety by integrating two existing tools [17], to cite just four examples.

Research in the use of computer algorithms to improve functional safety is prevalent, as in the case of computer-aided verification of functional safety [11, 12, 18]. Safety is critical in the aviation industry, which is always striving to develop improved ways to assess and predict risks, using techniques such as Bayesian Belief Networks [20]. One of our own prior lines of research tries to examine all the possible consequences of Off-Nominal Behaviors using a modeling technique called Casual Component Model (CCM) [13, 14]. Among the research areas, just cited, Hazard Space size is not directly addressed.

In summary, repository search using the term "Hazard Space" did not reveal prior publications in this area. Granted, there are many ways that a proposed solution to the hazard space problem could be expressed in a search engine, therefore making it difficult to find. There is also the possibility that "Hazard Space" is a term favored by this paper, but not by the majority of the FS community.

In any case, one can arguably conclude that published research in addressing the hazard space problem, as we have been defining it, is not as common as one might assume. This being the case, we cannot at this point cite prior work that directly relates to the problem addressed in this paper. A lack of hazard space related research could be due in part to reluctance in dealing with a concise hazard scenario. Scenarios that are long and detailed would tend to discourage the generation of every possible scenario due to large amount of project time that would have to be spend examining them. HSA sacrifices detail for the sake of hazard space coverage. We feel that once all the scenarios have been examined and risk assessed, the reminding relevant scenarios can be expanded to the desired level of detail.

Overview of Hazard Space Analysis (HSA)

We developed Hazard Space Analysis with several objectives in mind. First of all, we wanted a concise means of expressing a hazard scenario. Quite often in Functional Safety (FS) practice a harm sequence is manually written with a substantial degree of detail. This has its advantages, however, if the goal is to examine a system's entire hazard space, lengthy detailed harm sequences becomes impractical and can discourage stakeholders from generating a large number of them. A concise, one line, rule-based format, facilitates creating and examining the large number of scenarios inherent in a typical hazard space. Another advantage of HSA is that the rule-based format allows for each scenario to have a common structure and thus facilitates the automatic generation of hazard scenarios.

Finally, there is the ability to eventually map a given hazard scenario to a safety requirement; also made easier by a rule-based format, we call a safety rule. In this section we take a detailed look at how the notation used in HSA achieves these advantages. We begin by describing the format used in HSA's notation.

HSA Notation Format

A Hazard Scenario, as expressed in HSA's notation, uses a format similar to a function notation (F:A➜B). In this case the mapping symbol (➜) defines a transition from an Operational Situation (OS) to a Hazard State (HS) as caused (:) by some action or state (a.k.a. Hazard Cause). Thus a hazard scenario takes on the general form:

$$\text{Hazard Cause}(HC):\text{Operational Situation}(OS) \rightarrow \text{Hazard State}(HS) = HC:OS \rightarrow HS$$

The general form reads: The system transitions from an Operational Situation (OS) to a Hazard State (HS), when a Hazard Cause (HC) occurs. The notation further accounts for the fact that a hazard cause can be due to either a system Component Failure (CF), an Off-Nominal Behavior (ONB) exhibited by a human, a Nominal Behavior (NB) exhibited by a human or an Environmental State (ES). Thus, the four possible categorical variations of rules are:

- CF : OS ➔ HS, (i.e. A Component Failure causes a transition from Operational Situation to a Hazard State)

- ONB : OS ➔ HS, (i.e. An Off-Nom Behavior causes a transition from Operational Situation to a Hazard State)

- B : OS ➔ HS, (i.e. A Nominal Behavior causes a transition from Operational Situation to a Hazard State)

- ES : OS ➔ HS, (i.e. An Environmental State causes a transition from Operational Situation to a Hazard State)

An Operational Situation can be a nominal system state (such as a "vehicle moving forward"), whereas a Hazard State describes something potentially harmful to human, or damaging to property (such as "hit obstacle"). The following are four real world examples, according to hazard cause categories CF, ONB, NB, and ES:

- Brake failure : vehicle moving forward ➔ hit obstacle (a CF example)

- Operator turns off ignition : vehicle moving forward ➔ steering wheel locks up (a ONB example)

- Wet pavement : vehicle making sharp turn ➔ skids into ditch (an ES example)

- Makes sharp turn : vehicle moving forward ➔ Roll over (a NB example)

Each rule above describes a hazard scenario in a very concise manner. The emphasis is on what triggers a hazardous situation, and the trigger's precondition, in the form of an operational situation. The concise format allows for a direct mapping into a safety requirement, which specifies what needs to be addressed, not how it will be implemented. A risk ranking can be assigned to a hazard scenario.

A risk ranking is derived from two values: a severity value, as assigned to a hazard state and a probability of occurrence value as assigned to the hazard scenario, as a whole. The values can be arbitrary according to a predetermined scale. For example, severity can be ranked 1 to 5, with 5 being the most severe value. Probability can also be ranked 0 to 5, with 5 being the highest probability of occurrence and 0 meaning it will never occur or the scenario is irrelevant. Notation wise, the Severity value (Sv) of a Hazard State (HS), is indicated as HS(Sv), and the Probability of Occurrences (PO) value of a hazard rule, is indicated as {Hazard Scenario}(PO). For example if hitting an obstacle has a severity rating of 5, we would indicate that as (Hit Obstacle(5)). Furthermore, if a scenario containing a Hit Obstacle has a probability of 2, then that would be indicated by:

$$\{\text{Brake failure} : \text{vehicle moving forward} \to \text{hit obstacle}(5)\}(2) \qquad (2)$$

The risk ranking is calculated by multiplying Sv times PO, thus in R1 the hazard rule has a risk ranking of 5 x 2 = 10. Aside from being concise, the HSA notation allows for the automatic generations of all possible rules, given the initial entry of rule elements: CF, ONB, NB, ES, OS, and HS. Entering the rule elements, as lists, results in a lesser cognitive load than thinking of all the possible combinations in the form of rules. In our approach we have the human do what a human does best, while letting the computer do what it does best.

Generation of Hazard Space

A challenge addressed in this paper is trying to determine all the possible scenarios that can occur. HSA Notation is designed to facilitate the automatic generation of all possible hazard scenarios given the sets of OS, HS(Sv), CF, ONB, NB, and ES. It is mentally easier to conceive the individual elements of OS, HS(Sv), CF, ONB, NB, and ES, and have the computer automatically generate the scenarios from those elements.

For example, if asked to list all the possible Operational Situations that an automobile can be in, the average person would have little trouble listing entries such as: {Moving forward, Moving backward, Turning right, Turning left, etc.}. The same can be said when trying to assessing of a list of all possible Hazard States (e.g. {Hitting an obstacle, Skidding out of control, etc.}). Note that while conceiving of a hazard state, a severity value (Sv) of 1 to 5 can be assigned. (e.g. {Hitting an obstacle (5), skidding out of control (3), etc.}). Thus, if asked to compile a lists of all possible OS, HS(Sv), CF, ONB, NB, and ES, a person could merely focus on compiling each list individually, in particular, using the Functional Hazard Assessment defined in ARP 4761, to assess each hazard state.

Deriving all the combinations is something a computer can do very efficiently, once a human has compiled the individual lists of OS, HS(Sv), CF, ONB, NB, and ES. The following steps describe the generation of hazard scenarios which begins with step 1, the manual compiling of OS, HS(Sv), CF, ONB, NB, and ES.

Step 1: Manual compilation of OS, HS(Sv), CF, ONB, NB, and ES. This creates the following sets:

- $OS = \{os_1, ... os_n\}$ Set of Operational Situations
- $HS = \{hs(sv)_1, ... hsf(sv)_n\}$ Set of Hazard States
- $CF = \{cf_1, ... cf_n\}$ Set of Component Failures
- $ES = \{es_1, ... es_n\}$ Set of Environmental States
- $NB = \{nb_1, ... nb_n\}$ Set of human Nominal Behaviors
- $ONB = \{onb_1, ... onb_n\}$ Set of human Off-Nominal Behaviors

Step 2 is the automatic Cartesian product computation of the two sets *OS* and *HS(Sv)*, resulting in a set of ordered pairs consisting of *OS* members os_n and HS members $hs(sv)_n$:

$$(OS \times HS(Sv)) = \{os_1, ... os_n\} \times \{hs(sv)_1, ... hs(sv)_n\} = \{(os_1, hs(sv)_1), (os_1, hs(sv)_n), ... (os_n, hs(sv)_1), (os_n, hs(sv)_n)\}$$

Step 3 is the automatic translation of ordered pairs into transition statements, using the symbol (➜). This produces a set of transition statements, which we will represent with *(OS ➜ HS(Sv))*:

$$(OS ➜ HS(Sv)) = \{(os_1 ➜ hs(sv)_1), (os_1 ➜ hs(sv)_n), ... (os_n ➜ hs(sv)_1), (os_n ➜ hs(sv)_n)\}$$

Figure 2 shows a real world example of Steps 2 and 3 as it would apply to a fork lift.

Step 4 is the automatic Cartesian product computation of each of the sets compiled in Step 1 with the set of transitions computed in Step 3. For the sake of space, we will only expand the Cartesian product $CF \times (OS ➜ HS(Sv)) = \{(cf_1, os_1 ➜ hs(sv)_1), (cf_1, os_1 ➜ hs(sv)_n), (cf_1, os_n ➜ hs(sv)_1), (cf_1, os_n ➜ hs(sv)_n), ... (cf_n, os_1 ➜ hs(sv)_1), (cf_n, os_1 ➜ hs(sv)_n), (cf_n, os_n ➜ hs(Sv)_1), (cf_n, os_n ➜ hs(sv)_n)\}$. Figure 3 shows a real world example of Steps 4 and 5 as it would apply to a fork lift.

Step 5 is the automatic translation of ordered pairs in Step 4 into hazard scenarios, using the symbol (:). This produces a set of Hazard Scenarios, which we will represent

FIGURE 2   Real world example of Steps 2 and 3.

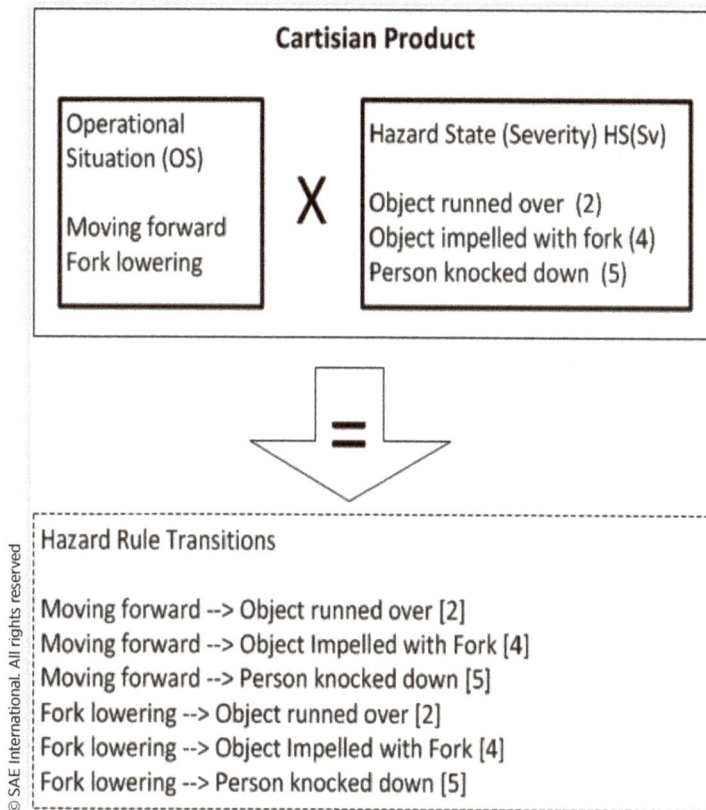

Cartisian Product

Operational Situation (OS)

Moving forward
Fork lowering

X

Hazard State (Severity) HS(Sv)

Object runned over (2)
Object impelled with fork (4)
Person knocked down (5)

=

Hazard Rule Transitions

Moving forward --> Object runned over [2]
Moving forward --> Object Impelled with Fork [4]
Moving forward --> Person knocked down [5]
Fork lowering --> Object runned over [2]
Fork lowering --> Object Impelled with Fork [4]
Fork lowering --> Person knocked down [5]

with $CF : OS \rightarrow HS(Sv) = \{(cf_1: os_1, \rightarrow hs(sv)_1), (cf_1: os_1 \rightarrow hs(sv)_n), (cf_1: os_n \rightarrow hs(sv)_1),$ $(cf_1: os_n \rightarrow hs(sv)_n), ... (cf_n: os_1 \rightarrow hs(sv)_1), (cf_n : os_1 \rightarrow hs(sv)_n), (cf_n :os_n \rightarrow hs(Sv)_1), (cf_n : os_n \rightarrow hs(sv)_n)\}$

Step 6: Applying Steps 4 and 5 to CF, ONB, NB, and ES produces the four categories of Hazard Scenarios:

$CF : OS \rightarrow HS(Sv)$, $ONB : OS \rightarrow HS(Sv)$, $NB : OS \rightarrow HS(Sv)$, and $ES:OS \rightarrow HS(Sv)$.

Steps 1 to 6 produces every single-caused combination of hazard cause, operational situation, and hazard state into a set of hazard scenarios. While every combination may not prove to be relevant, every possible combination can be generated for review. The engineer then reviews each rule's relevancy and assigns a Probability of Occurrence (PO) rating. An irrelevant rule is ignored by assigning a PO of 0, which when multiplied by the rule's Sv produces a risk ranking of 0. As an engineer/stakeholder rates each generated scenario, every possible hazard scenario can potentially be considered; the Hazard Space is reduced down to a relevant subset. This improves the chances that an important scenario is not overlooked since the engineer has the chance to examine a larger number of single-caused hazard scenarios. In aviation, there is a large potential for multi-caused hazard scenarios. These are scenarios that combine more than one cause, such as an Off-nominal behavior (ONB), resulting from a component failure (CF). The rule notation of HSA can accommodate a combination of hazard causes by logically ANDing two or more causes. Creating the entire hazard space of multi-caused scenarios would make a subtractive approach unfeasible. However, one possible approach is to focus on a

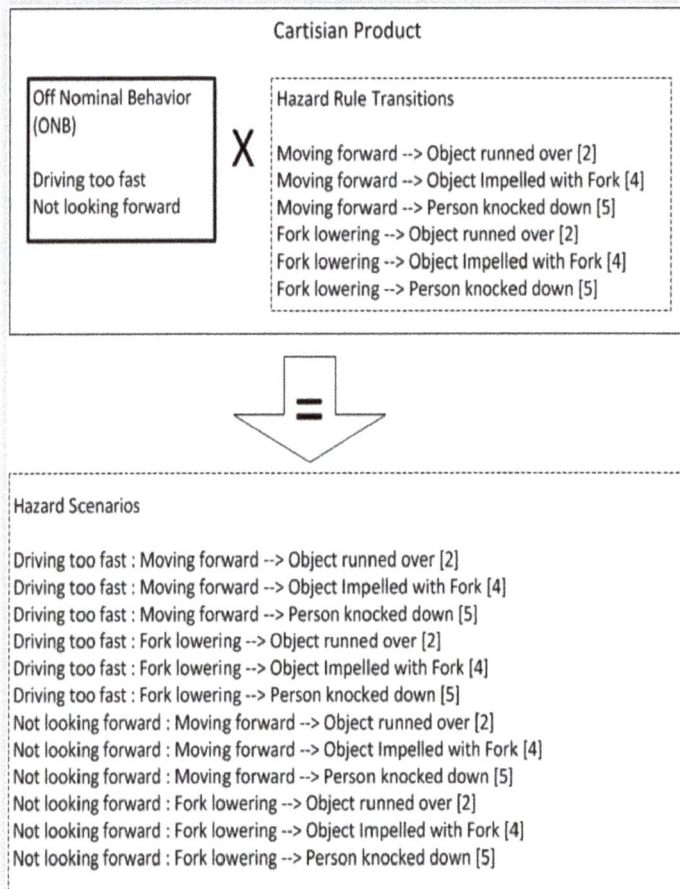

Cartisian Product

Off Nominal Behavior (ONB)

Driving too fast
Not looking forward

X

Hazard Rule Transitions

Moving forward --> Object runned over [2]
Moving forward --> Object Impelled with Fork [4]
Moving forward --> Person knocked down [5]
Fork lowering --> Object runned over [2]
Fork lowering --> Object Impelled with Fork [4]
Fork lowering --> Person knocked down [5]

=

Hazard Scenarios

Driving too fast : Moving forward --> Object runned over [2]
Driving too fast : Moving forward --> Object Impelled with Fork [4]
Driving too fast : Moving forward --> Person knocked down [5]
Driving too fast : Fork lowering --> Object runned over [2]
Driving too fast : Fork lowering --> Object Impelled with Fork [4]
Driving too fast : Fork lowering --> Person knocked down [5]
Not looking forward : Moving forward --> Object runned over [2]
Not looking forward : Moving forward --> Object Impelled with Fork [4]
Not looking forward : Moving forward --> Person knocked down [5]
Not looking forward : Fork lowering --> Object runned over [2]
Not looking forward : Fork lowering --> Object Impelled with Fork [4]
Not looking forward : Fork lowering --> Person knocked down [5]

single-caused hazard space, reduce that space by isolating the hazard scenarios with the highest risk/severity values, and combine the reduced number of single-causes with another reduced set of hazard causes. For example, we could assess component failures with the highest risk severity, narrow down to one component, and then use that one component in combination with all or a reduced set of ONB causes. This partitioning and combining of single-caused scenarios could mitigate the large hazard space produced by multi-caused scenarios. The partitioning and subsequent combining could be achieved by the computer, as the user selects which hazard causes to combine. We now look at how partitioning can mitigate a large hazard space of single-caused hazard scenarios.

Partitioning of a Single-Caused Hazard Space

Figure 4 is a graphical representation of steps 1 to 6; it shows how the hazard space is constructed from the sets of OS, HS(Sv), NB, ONB, ES, and CF. Where the sets intersect represent the Cartesian product of the two intersecting sets. For example, the intersection of OS and HS(Sv) contains OS ➔ HS(Sv), which is the Cartesian product of OS and HS(Sv). The hazard space is ultimately created by the combined intersection of {OS ➔ HS(Sv)} and NB, ONB, ES, and CF.

The size of the hazard space is defined by the following equation:

$$\left(|ONB|+|NB|+|CF|+|ES|\right)x\left(|OS|\,x\,|HS(Sv)|\right)$$
$$= Number\ of\ Hazard\ Scenarios \qquad (1)$$

|ONB| is the number of members in the set of ONB causes. The same applies to |NB|, |CF|, |ES|, |OS|, and |HS(Sv)|. Thus the number of transitions in {OS → HS(Sv)} is equal to the number of OS members times the number of HS(Sv) members. Similarly, for example, the number of hazard rules in {NB : OS → HS(Sv)} is equal to the number of NB members times the number of {OS → HS(Sv)} members. Equation (1) becomes the basis for partitioning the Hazard space into smaller, manageable, subsets by fixating one or more variable to 1. For example, let us assume that |ONB|, |NB|, |ES|, and |CF| each have the same number of members N. Let us also assume that |OS| and |HV(Sv)| each has M number of ‚members. Inserting values N and M into Equation 1, we get $(N + N + N + N) \times (M \times M) = 4N(M^2)$.

We now consider what happens if we partition the hazard space by fixating |ONB|, |NB|, |ES|, or |CF| to a value of one. In other words, suppose we want to examine the hazard space that corresponds to one member of ONB. The equation then becomes $(3N + 1)(M^2)$. If we fixate the value of |OS| or |HS(Sv)| to 1, the equation becomes $4N(M)$. If we fixate |OS| and |CF|, the equation becomes $(3N + 1)(M)$. Fixating both |ONB| and |NB| to one results in $(2N + 2)(M^2)$. There are obviously more possible combinations, which we will not explicitly show. The following shows the hazard space size for of the former examples, given that N = 3 and M = 4:

- The entire hazard space size: $(4N)(M^2) = 4(3)(4)^2 = 12(16) = 192$ hazard rules.

- Fixating |ONB|, |NB|, |ES|, or |CF| to 1 member: $(3N + 1)(M^2) = (3(3) + 1)(4)^2 = 10(16) = 160$ rules.

- Fixating |OS| or |HS(Sv)| to 1 member: $4N(M) = 4(3)(4) = 48$ rules.

- Fixating both |ONB| and |NB| to 1 member each: $(2N + 2)(M^2) = (2(3) + 2)(4^2) = 8(16) = 128$ rules

- Fixating both |OS| and |CF| to 1 member each: $(3N + 1)(M) = (3(3) + 1)(4) = 10(4) = 40$ rules.

Fixating |OS| or |HS(Sv)| results in a larger reduction of hazard space than fixating either |ONB|, |NB|, |ES|, or |CF|, something to take into account when partitioning a Hazard Space. Finally, there is also the possibility of partitioning based on the assigned Sv. This would allow focusing on a hazard subspace consisting of a desired severity level.

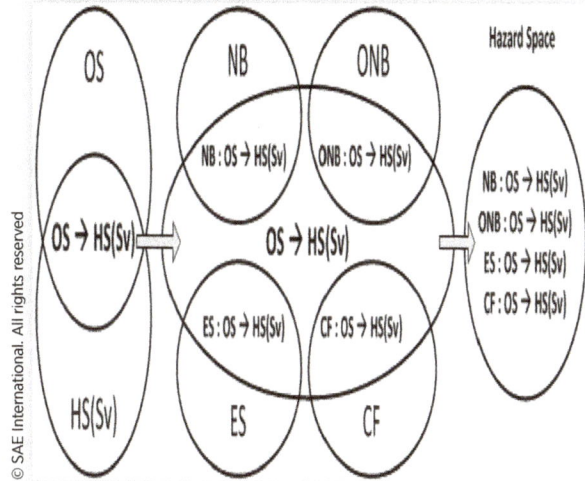

FIGURE 4 A diagram of the Hazard Space's construction.

Safety Rules from Hazard Scenarios

HSA notation enables one to algorithmically generate the hazard scenarios that make up the hazard space, and then selectively reduce the hazard space to those scenarios that are relevant and have a high risk rating. According to the IEC61508 standard, the ultimate goal is to develop a means of mitigating or eliminating the hazard scenarios that have been deemed high risk. This can be achieved by either integrating safety functions into the system being safe-guarded, and/or by developing a separate protection system that works in conjunction with the system being safe-guarded (refer back to Figure 1).

In either case, a safety function must be designed that addresses each given hazard scenario. Safety functions can be specified by safety requirements. In our case, the hazard scenario (rule) could be used to suggest a high level solution to hazard, in the form of safety rule. The details that better describe a safety requirement can be added later on. We now look at how a Hazard Scenario can be mapped directly to a safety rule by using the rule-based HSA notation. Since there should be a safety rule for each hazard scenario, the goal is to map each rule-based hazard scenario to a rule expressed with an IF, AND, THEN format. To illustrate how this mapping occurs, we refer back to the general definition of a Hazard Scenario (e.g. ES : OS → HS(Sv)). Recall that each scenario definition terminates in a Hazard State (HS) with a specified Severity value (Sv).

The goal is to specify a safety rule that counters, mitigates, or eliminates the HS term in a given hazard scenario. Thus, when defining a safety rule for a given scenario, we begin by looking at the scenario's HS term and conceive of an Anti-Hazard State. Figure 5 illustrates how a hazard scenario is mapped to a safety function. Figure 5A shows a general form of the scenario to function mapping. Figure 5B shows as example using a real world hazard scenario, namely, an automobile skidding out of control due to wet road conditions.

Note that the example, in Figure 5B, begins with determining a counter hazard state (in this case "Avoid skidding" counters "Skidding off road"). Once an Anti-HS is assessed, the next step is to determine what would tell the protection system to enact the Anti-HS. The diagram represents a future Safety Function as a safety rule that senses two

FIGURE 5 Hazard Scenario to Safety Rule.

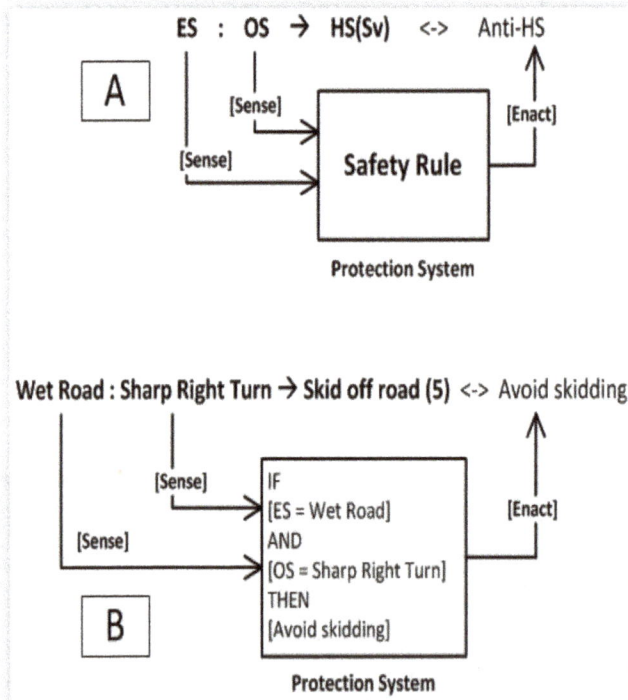

parameters (the Hazard Cause and OS) and then enacts a state/action that counters the given hazard state. The safety rule for the hazard scenario: "Wet Road : Sharp Right Turn → Skid of road (5)" becomes:

"IF [Wet Road] AND [Sharp Right Turn] THEN enact [Avoid skidding]."

If the safety functions are to be carried out by a separate protection system (running concurrently with the system being safe-guarded) the IF, AND, THEN format of the safety rule defines what the protection system needs to sense in order to enact the anti-HS. The rule's format also raises questions that must be answered, during the design of the protection system, to eventually produce a safety function. These questions may include the following:

1. Do we need a means to detect a slippery surface?
2. Do we need inertia/accelerometer sensors to determine when the car is turning?
3. How do we determine the car is slipping?
4. How do we prevent the car from slipping?

In fairness, we would not necessarily need to address a hazard scenario by a sense/enact approach. However, this format provides a systematic approach for addressing each hazard scenario. In the case where a safety function is to be directly integrated into the system being guarded, the safety rule's format can still be used to determine the combination of system state and inputs that must be sensed when reacting to a given hazard. The requirement's emphasis is on WHAT is involved in addressing a given hazard scenario. The HOW is addressed during the design of the safety function that comes from the safety requirement, initially defined by the safety rule.

There is, of course, the potential for a subsequent safety function's component to contribute to a previously non-existing hazard scenario; either by itself, or in combination with another hazard cause. One potential way to assess these additional hazards is to add the new safety component into the set of component-causes (CF) and reassess in a somewhat recursive manner. The other thing we should mention about safety rules and subsequent functions is that multiple anti-rules that address the same hazard scenario can be used to define redundant safety features, something that is often pursued in the aviation industry.

Aviation Safety Example

We now look at an example of applying Hazard Space Analysis to examine some of the potential hazards associated with flying an aircraft. Table 1 shows seven potential hazards (HS(Sv)) with severity levels (Sv) ranging from 1 to 5. The table also lists some potential Aircraft component failures (CF), some fight Environmental States (ES) that can be encountered by the pilot/aircraft, and some potential Off-Nominal Behaviors (ONB) by the pilot.

To keep the example's hazard space within a reasonable size, we will focus on three operational situations (OS) during the aircraft's airborne phase: level flying, ascending and descending. Using the HSA algorithm explained in the overview section, the table entries will be algorithmically combined to create a set of hazard scenarios. The hazard scenarios will then be risk assessed and used to derive some safety requirements.

Managing a Large Hazard Space

If we define Off-Nominal Behavior (ONB) as those behaviors that go contrary to how a pilot would have been trained to behave, we can ask "What are some of the Off-Nominal

TABLE 1 Elements of the Aviation Safety Example

Pilot (ONB)	Aircraft (CF)	Flight Environment (ES)	Situation (OS)	Hazard (HS(Sv))
No scanning	Stuck elevator surface	Night-time visibility	descending	Midair collision (5)
Rapid pull back of control wheel	Gyroscope failure	Wake turbulence	Level flying	Coriolis illusion (2)
Using Kinesthesis sense	Carburetor icing	Wing blind spot	ascending	Stalled engine (4)
Prolonged constant-rate turn	landing gear not deployed	Congested area		Somatogravic illusion (4)
	Landing lights failure	Micro bursts		Empty field myopia (3)
				Autokinesis (2)
				Stall (3)

Behaviors that a pilot can exhibit"? It would not be hard for an experienced aviator to come up with some of the items listed under the column entitled Pilot (ONB). Table 1 lists "No scanning" as a pilot's Off-Nominal Behavior (ONB). This behavior goes contrary to the recommended pilot practice of continuously scanning with his/her eyes and not fixating on any given spot in the field of the vision. The ONB list is not exhaustive, but sufficient for our example.

As to the list of hazards (HS(Sv)) in Table 1, we focused on hazards that could lead to serious harm if not kept in check.

The generation of the hazard space, from Table 1, is shown in Figure 6 (represented in a similar manner as Figure 4). Note that a severity level (on a scale of 1 to 5) has been assigned to each hazard. We rated "Midair collision" with the highest severity of 5, and a "stalled engine" with a 4 (in the case of a single prop plane, a stalled engine could be a 5). A wing "Stall" is considered less severe, since it can typically result in a temporary drop in altitude, which can often be recovered from. The erroneous sensation of rotation known as Coriolis illusion is rated 2, since the typical pilot compensation should not result in a fatal catastrophe in a more experienced pilot.

We rated Somatogravic illusion as more severe at 4, since the illusion that the nose of the plane is higher than it really is can result in a dangerous nose dive. Empty field myopia, with its effect on focusing, can interface with a pilot's avoidance of a mid-air collision, so we rated it at 3.

Finally, we rated Autokinesis at 2 because at worst a pilot could get off course while flying at night, due to mistaking a stationary light as one that is moving. Assigning severity levels can be subjective. We strongly recommend that these levels be assigned by domain experts and/or stakeholders that are well familiar with the hazards in question. The two additional sets of hazard causes are Aircraft component failure (CF) and Flight Environment (ES). With the three values for OS, the size of the hazard space is calculated at $(4 + 5 + 5)(3 \times 7) = 14(21) = 294$; a considerable number of hazard scenarios to examine. Recall that a hazard space can be partitioned by fixating ONB, CF, ES, OS or HS(Sv). So we can further confine the hazard space inspection to "level flying", which yields a hazard space size of $(4 + 5 + 5)(1 \times 7) = 14(7) = 98$. A further size reduction can also be achieved by ignoring one or more hazard causes. In this case we will ignore CF and ES and focus on ONB. This yields a size of $(4 + 0 + 0)(1 \times 7) = 4(7) = 28$.

FIGURE 6 Creation of partitioned Hazard Space for Aviation example.

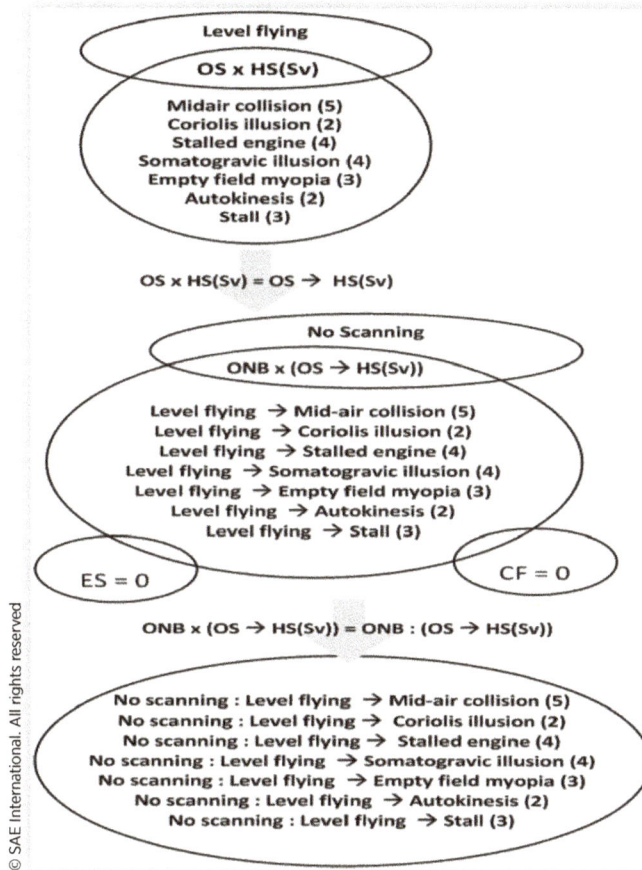

Level flying

OS x HS(Sv)

Midair collision (5)
Coriolis illusion (2)
Stalled engine (4)
Somatogravic illusion (4)
Empty field myopia (3)
Autokinesis (2)
Stall (3)

$$OS \times HS(Sv) = OS \rightarrow HS(Sv)$$

No Scanning

$$ONB \times (OS \rightarrow HS(Sv))$$

Level flying → Mid-air collision (5)
Level flying → Coriolis illusion (2)
Level flying → Stalled engine (4)
Level flying → Somatogravic illusion (4)
Level flying → Empty field myopia (3)
Level flying → Autokinesis (2)
Level flying → Stall (3)

ES = 0   CF = 0

$$ONB \times (OS \rightarrow HS(Sv)) = ONB : (OS \rightarrow HS(Sv))$$

No scanning : Level flying → Mid-air collision (5)
No scanning : Level flying → Coriolis illusion (2)
No scanning : Level flying → Stalled engine (4)
No scanning : Level flying → Somatogravic illusion (4)
No scanning : Level flying → Empty field myopia (3)
No scanning : Level flying → Autokinesis (2)
No scanning : Level flying → Stall (3)

Risk Assessment

Starting with the just partitioned 28 hazard scenarios, we will focus on all the scenarios caused by "No scanning". This brings the final number of scenarios, being examined, to seven, as shown in the oval at the bottom of Figure 6. These seven scenarios are:

1. No scanning : Level flying → Mid-air collision (5)
2. No scanning : Level flying → Coriolis illusion (2)
3. No scanning : Level flying → Stalled engine (4)
4. No scanning : Level flying → Somatogravic illusion (4)
5. No scanning : Level flying → Empty field myopia (3)
6. No scanning : Level flying → Autokinesis (2)
7. No scanning : Level flying → Stall (3)

As engineers and/or stakeholders review the seven scenarios listed above, they are to assess the probability of each scenario occurring according to whatever systematic means they have at their disposal. If the scenario is deemed irrelevant, then a probability of zero is assigned. We use a scale of 0 to 5, with 5 being the highest probability of occurrence. Starting with scenario number 1, we determine that fixating on one position or object can result in a midair collision, especially when one considers the combined speed of the two aircrafts involved. However, considering that there are FAA regulations and safety systems that would minimize a collision, we rate scenario 1 with a probability of 4. In Scenario 2, Coriolis illusion is typically not assorted with lack of visual scanning, we rate scenario 2 with a probability of 0, which in essence, means the scenario is irrelevant.

Observe that even though scenario number 2 is considered irrelevant, it is a member of the hazard space, and therefore became subject to review. It could have been the case that there is a correlation between the lack of scanning and the Coriolis illusion. Conventional wisdom may say not likely, but perhaps one of the stakeholders can cite an example of such a correlation. In any case, the scenario would likely never occur to an engineer, had the hazard space been compiled in an additive manner, whereas in our subtractive approach scenario 2 was generated and reviewed.

Scenario number 3 can initially be deemed irrelevant, but again brings to mind whether a correlation between not visually scanning and a stalled engine is possible. Having brought this scenario to mind, perhaps one of the stakeholders will conceive of a pilot not noticing a very low fuel level, due to not readily scanning the cockpit instrumentation. So what initially starts as a probability of 0, might become a slight possibility of 1 or 2, and may prompt the design of a new safety feature.

Scenarios 4 and 7 are not readily relevant. Whereas, scenarios 5 and 6 describe hazards that can occur due to staring, so they rate a probability of 5 each. The question again occurs whether scenarios 5 and 6 would have been conceived using an additive approach, considering the combinations of situations, hazards, and pilot behavior. After assigning a probability to each scenario and calculating a risk factor, we get:

{No scanning : Level flying ➜ Mid-air collision (5)} (4) = 20

{No scanning : Level flying ➜ Coriolis illusion (2)} (0) = 0

{No scanning : Level flying ➜ Stalled engine (4)} (2) = 8

{No scanning : Level flying ➜ Somatogravic illusion (4)} (0) = 0

{No scanning : Level flying ➜ Empty field myopia (3)} (5) = 15

{No scanning : Level flying ➜ Autokinesis (2)} (5) = 10

{No scanning : Level flying ➜ Stall (3)} (0) = 0

From the risk assessed scenarios we determine that the highest risk, rated at 20, are the possibility of a mid-air collision. The next highest is Empty field myopia, at 15, which can also result in mid-air collision. The next highest risk is flying off course at night due to Autokinesis, at 10. The lowest risk (at 8) is the possibility of a stalled engine due to no scanning, which may not have occurred to anyone trying to imagine that hazard scenario off the top of their head.

Finally we should note that the hazard scenarios with a risk rating of 0, can be informative as well. For example, suppose that after a complete set of hazard scenarios are examined, we notice that none of the hazards associated with Somatgravic illusion ever have a risk rating higher than zero. The initial conclusion may be that Somatgravic illusion is a non-issue, however the question should be whether the cause of Somatgravic illusion has been considered part of the Pilot's potential off-nominal behavior; perhaps it should have. For example, we might conclude that a rapid acceleration of the aircraft, which is a common cause of Somatgravic illusion, was overlooked as a potential Pilot ONB. In this case, the assessed irrelevancy of Somatgravic illusion led to the discovery of an overlooked ONB candidate. This illustrates another advantage of exploring as much of the hazard space as possible, namely the potential for the discovery of overlooked hazards, causes, and operational situations.

Safety Requirements

We will now look at how a set of safety requirements can be derived from the hazard scenarios assess with higher risk ratings. We will use the four scenarios from our on-going example:

{No scanning : Level flying ➔ Mid-air collision (5)} (4) = 20

{No scanning : Level flying ➔ Empty field myopia (3)} (5) = 15

{No scanning : Level flying ➔ Autokinesis (2)} (5) = 10

{No scanning : Level flying ➔ Stalled engine (4)} (2) = 8

Translating a hazard scenario to a safety requirement begins by imagining a safety function that uses two inputs as triggers that enacts an anti-hazard state (refer back to Figure 5). The two inputs are triggered by the hazard scenario's hazard cause (HC) and operational situation (OS). In the case of the four scenarios listed above, HC is "No scanning" and OS is "level flying" for all four scenarios. This means that a protection system would have to sense a plane at level fight. This would not be difficult to achieve, since planes already come equipped with attitude indicators as part of their instrumentation system. Sensing the lack of visual scanning may involve developing an eye tracking device for pilots, or some other creative means. A means of independently verifying a pilot's scanning habits can be applied to the four hazard scenarios in question. The anti-HS would be potentially different for each scenario, however, with some creative thought it may be possible to conceive of one or two anti-HS that addresses all four scenarios. One common solution might be a cockpit alarm triggered by lack of pilot scanning during level flying.

Discussion and Limitations

Three arguable limitations to HSA are the potentially large number of hazard scenarios to be examined, the conciseness of the hazard scenarios, and the fact that the hazard space size generated is limited by the number of CF, ONB, NB, ES, HS, and OS that can be humanly conceived. The question of examining a large hazard space was addressed during our aviation example when we partitioned the hazard space by focusing on only hazards caused by the pilot's off-nominal behaviors. Note that we could have partitioned from the start by eliciting only ONB, however, it is still important to elicit all the potential hazard cause categories (such as ONB, NB, CF, and ES), during the requirements elicitation phase. Recall, that functional safety is not an afterthought, but a process that occurs concurrently with the development of the system being safeguarded. The same stakeholders (and domain expects) familiar with what the system needs to do, are typically also well-versed in what can go wrong with the system. Therefore, it is best to gather all the potential hazard causes up-front, and then partition as needed.

As to the conciseness question, the conciseness of a hazard scenario serves as an approximation to the scenario being addressed; details can always be later added, during the design phase. The goal with keeping the scenarios concise is to facilitate the generation of a large hazard space. Once the hazard space has been reduced to those scenarios that are relevant and present a high risk, those scenarios can be expanded into greater detail, particularly when creating the safety functions. Thus, hazard scenarios are concise to facilitate the exploration of a large hazard space.

Hazard space size generation is dictated by the number of elements known as hazard causes, operational situations, and hazard state that can be conceived by the engineer/stakeholders. This means that if a key element is overlooked, the hazard space size is not as large than it could be. We don't view this as necessarily problematic because an additive approach, where stakeholders would be incrementally conceiving the scenarios, would be confronted with the same problem, but with a greater risk of missing key scenarios. Therefore, our subtractive approach is still providing an advantage, even if elements are sometimes overlooked. To minimize the problem of overlooked members of ES, ONB, NB, HS, OS, and CF, we recommend that domain experts be part of the process. Particularly those experts that are familiar with how the system being designed will interact with its operating environment. Such interactions are where a lot of the potential hazards can occur.

One area we already discussed in some detail is the idea of combining hazard causes (the generation of hazard space section); for example, a scenario that combines an operator off-nominal behavior with an environment state, such as, driving too fast (ONB) while the road is wet (ES). Users of HSA are free to logically combine hazard causes, however, as mentioned, we recommend that combinations be performed after all the Cartesian products have been first calculated with a single hazard cause; perhaps even after risk assessments have been performed. With multiple hazard causes combined, translating into a safety rule would then involve three or more terms as triggers (instead of just one hazard cause and OS). There is also the possibly of using GPUs (instead of the slower CPUs) to manage a multi-caused hazard scenario, without having to first reduce a set of single-caused rules. The use of GPUs has help mitigate the node explosion associated with deep-learning neural nets, perhaps they could also manage the large number of hazard rules associated with the hazard space typically encountered in the aviation industry. A combination of GPUs and partitioning might also make a subtractive approach to examining a hazard space, more feasible. In the meantime, we feel that this subtractive approach can still be applied to a number of problems in software engineering where the goal is to analyze a number of candidates whose number is closer to the maximum search space size, whenever the search space is manageable.

Another area to explore is the possibility of expanding the categories of hazard causes, such as adding an operator's unexpected health event, or the intentional tampering of safety features. The categories can even be expanded to model security related hazards by including a cyber-based hazard cause, such as, a Hacking Event (HS). Finally, one theoretical implication of this work is the idea of expanding a search space to its maximum size and then reducing it to the desired subspace

Conclusion

Our goal in this paper is to present a potential means of examining a system's entire hazard space, so as to prevent overlooking a potentially high-risk hazard. This has resulted in the following contributions to the area of safety assessment. First of all, there is the idea of using a subtractive approach to generate relevant hazard scenarios. The more typical approach of adding scenarios as they are assessed by engineers and stakeholders, has certain limitations. We felt that history has examples of system failures that could have been anticipated, had the designer "considered that particular scenario." We also considered whether most if not all of a hazard space can be examined, providing we sacrificed scenario detail for space size, by using a rule-based approach. The idea was to first assess as many scenarios as possible, and then add detail to those scenarios that remain.

The second contribution is using a rule based format that lends itself to the algorithmic generation of the hazard space. This allows for a computer to generate the combinations of operational situations, hazards, and causes, rather than leave that generation up to the human engineer and stakeholders. The rule-based format also allows for a direct mapping of a hazard scenario to a safety rule, which defines what a safety function is supposed to do. Future directions for this research can include automated analysis of the hazard scenarios by looking for correlations between high percentages of a given hazard state and hazard cause. Cluster analysis could also potentially show patterns in the scenarios that could enable fewer safety functions to address a large percentage of hazards.

In summary, while there is always the question of how to deal with a very large hazard space, HSA can serve as a framework to potentially raise questions about large portions of the hazard space, if we apply even more creative means of partitioning, selecting, and combining hazard scenarios in a very concise, rule-based manner.

Contact Information

The author can be contacted via
DISTek Integration, Inc
daniel.aceituna@distek.com

References

1. Bernardini, A., Ecker, W., and Schlichtmann, U., "Where Formal Verification Can Help in Functional Safety Analysis," *Paper presented at the Meeting of the ICCAD*, 2016.

2. Sieker, B.M., "A Proposal for Improving the Applicability of Formal Methods in the Functional Safety Base Standard IEC.61508," *System Safety and Cyber-Security Conference*, 2015, doi:10.1049/cp.2015.0279.

3. Bernardini, A., Ecker, W., and Schlichtmann, U., "Efficient Handling of the Fault Space in Functional Safety Analysis Utilizing Formal Methods," *Paper presented at the Meeting of the VLSI-SoC*, 2016.

4. Allenby, K. and Kelly, T., "Deriving Safety Requirements Using Scenarios," *Paper presented at the Meeting of the RE*, 2001.

5. George, A., Taylor, W., and Nelson, J., "Writing Good Technical Safety Requirements," SAE Technical Paper 2016-01-0127, 2016, doi:10.4271/2016-01-0127.

6. George, A. and Nelson, J., "Managing Functional Safety (ISO26262) in Projects," SAE Technical Paper 2017-01-0064, 2017, doi:10.4271/2017-01-0064.

7. Putz, M., Seifert, H., Zach, M., and Peternel, J., "Functional Safety (ASIL-D) for an Electro Mechanical Brake," SAE Technical Paper 2016-01-1953, 2016, doi:10.4271/2016-01-1953.

8. Moure, C. and Kersting, K., "Development of Functional Safety in a Multi-Motor Control System for Electric Vehicles," SAE Technical Paper 2012-01-0028, 2012, doi:10.4271/2012-01-0028.

9. Krithivasan, G., Taylor, W., and Nelson, J., "Developing Functional Safety Requirements Using Process Model Variables," SAE Technical Paper 2015-01-0275, 2015, doi:10.4271/2015-01-0275.

10. Brewerton, S., "A New Approach to Input and Output Monitoring for Microcontrollers Supporting Functional Safety," *SAE Int. J. Passeng. Cars - Electron. Electr. Syst.* 6, no. 1 (2013): 126-133, doi:10.4271/2013-01-0185.

11. Bernardini, A., Ecker, W., and Schlichtmann, U., "Where Formal Verification Can Help in Functional Safety Analysis," *Paper presented at the Meeting of the ICCAD,* 2016.

12. Mader, R., Griessnig, G., Leitner, A., Kreiner, C., Bourrouilh, Q., Armengaud, E., Steger, C., and Weiß, R., "A Computer-Aided Approach to Preliminary Hazard Analysis for Automotive Embedded Systems," *Paper presented at the Meeting of the ECBS,* 2011.

13. Aceituna, D., "Elicitation Practices That Can Decrease Vulnerability to Off-Nominal Behaviors: Lessons from Using the Causal Component Model," *SAE Int. J. Passeng. Cars - Electron. Electr. Syst.* 10, no. 1 (2017): 83-94, doi:10.4271/2016-01-8109.

14. Aceituna, D. and Do, H., "Exposing the Usceptibility of Off-Nominal Behaviors in Reactive System Requirements," *Paper presented at the Meeting of the RE,* 2015.

15. Denney, E., Ganesh, P., and Ibrahim, H., "Perspectives on Software Safety Case Development for Unmanned Aircraft," *Dependable Systems and Networks (DSN), 2012 42nd Annual IEEE/IFIP International Conference on,* IEEE, 2012, 1-8.

16. Chen, D., Rolf, J., Henrik, L., Yiannis, P., Anders, S., Fredrik, T., and Martin, T., "Modelling Support for Design of Safety-Critical Automotive Embedded Systems," *International Conference on Computer Safety, Reliability, and Security,* Springer Berlin Heidelberg, 2008, 72-85.

17. Schulze, M., Jan, M., and Danilo, B., "Functional Safety and Variability: Can It be Brought Together?," *Proceedings of the 17th International Software Product Line Conference,* ACM, 2013, 236-243.

18. Täubig, H., Udo, F., Christoph, H., Christoph, L., Stefan, M., Elena, V., and Dennis, W., "Guaranteeing Functional Safety: Design for Provability and Computer-Aided Verification," *Autonomous Robots* 32, no. 3 (2012): 303-331.

19. Denney, E. and Ganesh, P., "Automating the Assembly of Aviation Safety Cases," *IEEE Transactions on Reliability* 63, no. 4 (2014): 830-849.

20. Luxhoj, J.T., Probabilistic Causal Analysis for System Safety Risk Assessments in Commercial Air Transport," 2003.

21. Bell, R., "Introduction to IEC 61508," *Proceedings of the 10th Australian Workshop on Safety Critical Systems and Software,* Australian Computer Society, Inc, 2006, vol. 55, 3-12.

3

A Model-Driven Approach for Dependent Failure Analysis in Consideration of Multicore Processors Using Modified EAST-ADL

Bülent Sari
ZF Friedrichshafen AG

Hans-Christian Reuss
University of Stuttgart/FKFS

Safety is becoming more and more important with the ever increasing level of safety related E/E Systems built into the cars. Increasing functionality of vehicle systems through electrification of power train and autonomous driving leads to complexity in designing system, hardware, software and safety architecture. The application of multicore processors in the automotive industry is becoming necessary because of the needs for more processing power, more memory and higher safety requirements. Therefore it is necessary to investigate the safety solutions particularly for Automotive Safety Integrity Level (ASIL-D) Systems. This brings additional challenges because of additional requirements of ISO 26262 for ASIL-D safety concepts. This paper presents an approach for model-based "dependent failure analysis" which is required from ISO 26262 for ASIL-D safety concepts with decomposition approach. Therefore, the hardware modeling, function modeling and dependability package of EAST-ADL (Electronics Architecture and Software Technology - Architecture Description Language) are extended in a way that it now allows the modeling of a multicore processor with its hardware elements and software safety architecture which are necessary to prove hardware and software independency. Additionally, some scripts are developed to analyze the decomposition paths automatically

CITATION: Sari, B. and Reuss, H., "A Model-Driven Approach for Dependent Failure Analysis in Consideration of Multicore Processors Using Modified EAST-ADL," SAE Technical Paper 2017-01-0065, 2017, doi:10.4271/2017-01-0065.

from system level to software and hardware level and generate the analysis results. Additionally, we briefly discuss how the main activities from ISO 26262 such as hazard analysis and risk assessment, functional safety concept, technical safety concept, safety analysis, etc. can be developed model-driven. The extensions and developed scripts make it possible to gain sufficient transparency and traceability for the safety arguments and to support the whole safety process in a single solution even in hardware and software development.

Introduction

Nowadays, in a premium vehicle up to 100 processing units (ECUs) are installed, which are capable of computing complex algorithms. Overall, the embedded software of a premium automobile contains up to 100 million lines of source code. On the contrary, new "Boeing Dreamliner 787" needs for all onboard systems around 6.5 million lines of code [1].

This comparison shows how complex the software in today's vehicles. This complexity and the scale of the software will continue growing [1]. The reason for the large amount of software requirement is the electrification of the automobile and autonomous driving systems. It is believed that the ratio of electronic components to the total production cost of a vehicle could rise up to 35% by 2020 and up to 50% by 2030 [2]. With the increase of electrification, the proportion of safety-critical systems is also grown. The malfunctions as "unintended blocking of the drive axle while driving", "power assisted steering acts in the wrong direction", and "false tripping of the airbag while driving" are a few examples that can lead to life-threatening injuries [3] and are therefore classified with ASIL-D.

The ISO 26262 provides the possibility to apply decomposition approach for ASIL-D safety requirements. An appropriate decomposition has the advantage to reduce the ASIL rating of the top events. But the application of ASIL decomposition requires redundancy of safety requirements, which should be allocated to sufficiently independent architectural elements. In order to apply the decomposition, ISO 26262 requires to prove "freedom from interference (FFI)" and to carry out "dependent failure analysis (DFA)", which then provides evidence about sufficient independency between decomposed function parts. Currently, it is possible to use commercial external tools [10, 11] to perform only the safety analysis within EAST-ADL. These tools are capable to generate fault tree analysis (FTA) and failure modes and effects analysis (FMEA) automatically from an EAST-ADL error model. But dependent failure analysis is realized currently manually and there is no supporting tool for that. This causes additional development effort, because the whole path from system decomposition down to software and hardware decompositions has to be analyzed to ensure that the signals and hardware parts are sufficiently independent. The other disadvantage of manual analysis is that it is difficult to achieve traceability.

An approach, how the engineers deal with these challenges, uses model-driven system, software and safety development. This makes it possible to describe, analyze and verify the system, software and safety architecture with models in order to detect the design and systematic errors before the implementation or code generation. Due to the high complexity of E/E-Systems in the vehicle, the creation of the system and software architecture is divided into different abstraction levels. This allows that the design to be checked on each level, which leads to a very early evaluation of the system architecture based on functional and non-functional requirements [4].

The modeling of a complete system is an extensive project that requires domain knowledge. In addition, knowledge is needed about various tools, because the architecture is described using several tools. Alternative to the current approach is a domain-specific language based on an ADL [5]. So it is possible that all information is created within a system model that describes the complete system, safety and software architecture.

The extensions and developed scripts make it possible to gain sufficient transparency and traceability for the safety arguments in consideration of multicore processors and to support the whole safety process in a single solution even in hardware and software development.

Description of the Approach

Approach of System and Safety Modeling [12]

The approach in Figure 1 shows how the system, software and safety development are merged, and thus the complexity of the system is reduced by a continuous architecture.

The developed method consists of five main parts, which will be shown in Figure 2 and the details will be explained below:

1. First part is the architecture development [6]. This part is about the creation of feature model, system analysis, functional design architecture (FDA) and hardware design architecture (HDA). It is also possible to allocate the functions from FDA model to the corresponding hardware elements of HDA model. HDA-Model is developed further to enable the modeling of multicore processors with additional hardware elements.

FIGURE 1 Model driven approach for system, safety and software development.

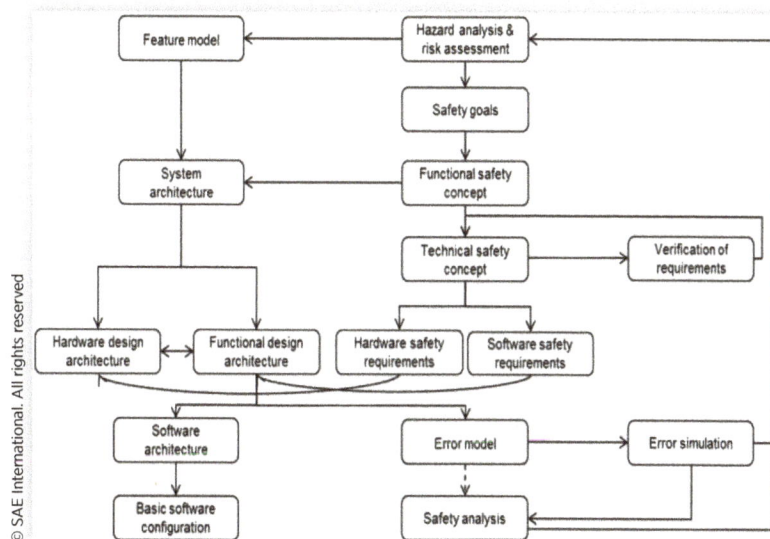

FIGURE 2 The main parts of the approach.

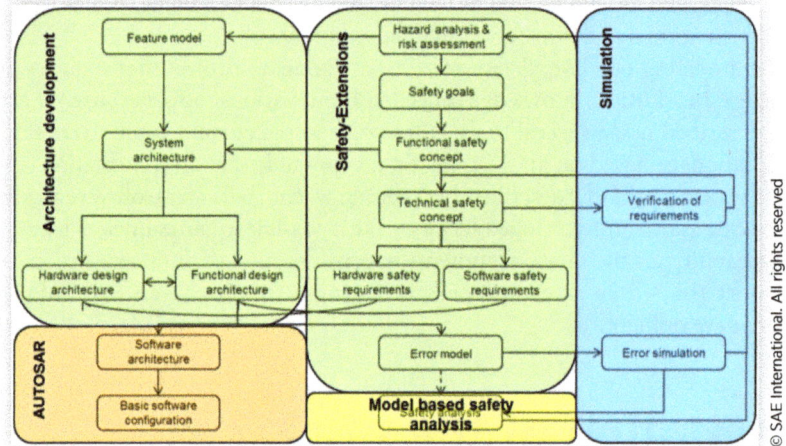

2. Second part is safety extensions [7] [8]. This part deals with the model-based creation of ISO 26262 work products. Firstly, hazard analysis and risk assessment is performed. Secondly, the safety goals are derived from hazard analysis and risk assessment. In order to fulfill the safety goals, functional safety concept and technical safety concept are created in the following step. Feature model can contain both non-safety-critical and safety-critical properties of the system. After hazard analysis and risk assessment, the safety-critical aspects of the system will be considered into the feature model.

   In the system architecture, the safety goals and corresponding safety functions of the system are taken into account. In the functional and hardware design architecture, the safety-critical functions are detailed further.

3. Third part is AUTOSAR [9]. This is about the software architecture and basic software configuration. The software architecture can be created from FDA model, which should contain the necessary software features. The implementation level is developed further to enable the signal allocation with hardware elements.

4. Fourth part is model-based safety analysis. In this step, the fault trees (FTA) of fault models are automatically generated from error model of ADL or simulation model with external tools [10] [11]. This part is investigated further to develop a new approach about the model based dependent failure analysis (DFA), which is described in the section B.

5. The last part is simulation and verification. In this step, error simulation and verification of the requirements are carried out in the earlier phase of development. On the one hand, the safety requirements are verified with usage of a simulation environment. On the other hand, it is possible to determine the system impacts of causes with error simulation. Thus, the defined top events of the system from hazard analysis and risk assessment can be approved. The possibility of the simulation and verification within EAST-ADL will be investigated within a separate paper.

The method allows for consistency and traceability of individual steps. The method also permits the efficient tracing from the software architecture to the feature model and from the safety analysis to the hazard analysis and risk assessment.

The approach that is presented in this paper has the following advantages:

- Modeling of the safety-related functions of an electric vehicle in an architecture description language.

- Achievement of an efficient and consistent model-based development of automotive embedded systems.

- Model based creation of ISO 26262 work products from hazard analysis and risk assessment to safety requirements.

- Merge of system and safety development.

- Early detection of systematic errors and system impact of errors.

- Model driven approach for dependent failure analysis

The details of the main parts have been described in another paper [12].

Approach of DFA-Analysis

As mentioned above, the ISO 26262 provides the possibility to apply decomposition approach for ASIL-D safety requirements. An appropriate decomposition has the advantage to reduce the ASIL rating of the top events. But the application of ASIL decomposition requires redundancy of safety requirements, which should be allocated to sufficiently independent architectural elements.

ISO 26262 mentions the following requirements to the decomposition approach [13]:

"As a basic rule, the application of ASIL decomposition requires redundancy of safety requirements allocated to architectural elements that are sufficiently independent."

"If the architectural elements are not sufficiently independent, then the redundant requirements and the architectural elements inherit the initial ASIL."

"In the case of use of homogenous redundancy (e.g. by duplicated device or duplicated software) and with respect to systematic failures of hardware and software, the ASIL cannot be reduced unless an analysis of dependent failures (see clause 7) provides evidence that sufficient independence (see ISO 26262-1:2018,1.74) exists or that the potential common causes lead to a safe state. Therefore, homogenous redundancy is in general not sufficient for reducing ASIL due to the lack of independence between the elements."

As shown in Figure 3, the ASIL of safety goal is inherited by corresponding safety requirements and safety functions. Figure 3 shows additionally, that the decomposed functions of sufficiently independent architectural elements inherit the original ASIL information in the brackets.

The purpose of dependent failures analysis is to find out single causes that could prevent to fulfill a required independence or freedom from interference requirements between architectural elements and violate a safety requirement. As given in ISO 26262-9, independence of architectural elements is threatened by common cause failures and cascading failures, while freedom from interference is only violated by cascading failures.

This research focused on the analysis of architectural features such as similar redundant elements, partitions of functions or software elements, shared resources and the used signals. The developed model based DFA-Analysis is able to check whether the

FIGURE 3   ASIL-Decomposition within EAST-ADL.

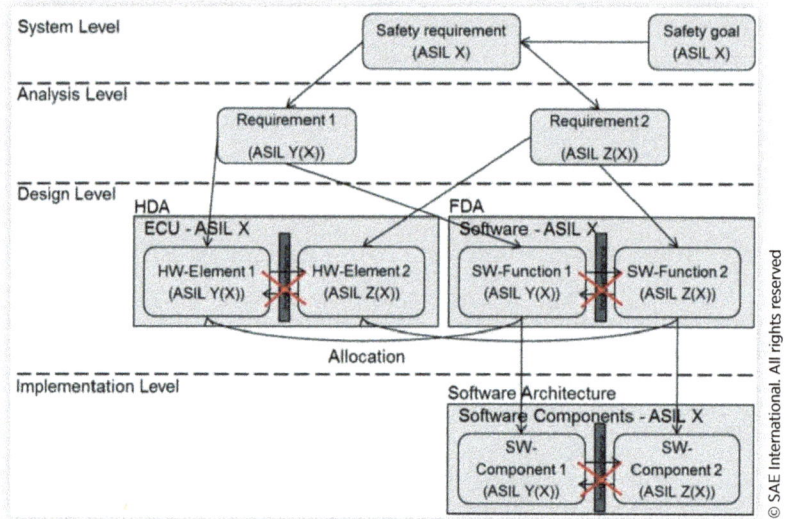

evidence of sufficient independence between the decomposition paths and signals exist. The details of the developed approach are described in the following subsections.

A. NECESSARY DEVELOPMENTS OF EAST-ADL FOR THE DFA ANALYSIS

The purpose of EAST-ADL is to provide the engineers facilitation for the representation and description of electronic systems in vehicles in a standardized form [4]. It can be used for different activities as modeling of functional requirements, safety work products from ISO 26262 as well as analysis and design purposes [6].

As shown in Figure 3 and Figure 4, the metamodel of EAST-ADL is organized into four different abstraction levels "system level, analysis level, design level and implementation level". Each of them fulfills specific roles. Each level considers a different phase of vehicle development and provides different perspectives of the whole system architecture. But currently the safety aspect is not considered by the modeling of these

FIGURE 4   ASIL-Decomposition with extended attributes.

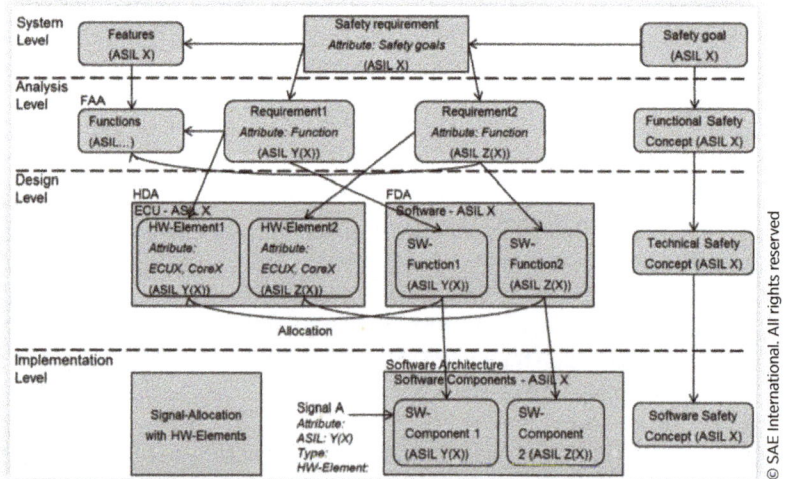

abstraction levels. The safety modeling is applied separately with the dependability package of EAST-ADL where the ISO work products such as hazard analysis and risk assessment, safety goals and safety concepts can be modeled. But it is necessary to extend the library elements of EAST-ADL with additional safety features which are used for the architecture modeling from the abstraction levels (FAA, FDA, HAD, etc.), to develop an automated dependent failure analysis approach.

Figure 4 shows the extended attributes such as ASIL, type of signals and hardware elements:

In addition to the new attributes for the existing library elements, additional HDA Library elements (Hardware Design Architecture) should be developed. Currently, hardware design modeling is kept abstract and consists of the following elements:

"sensor, actuator, node (ECU), power supply, hardware component prototype, logical bus, IOHardwarePin, communication hardware pin, power hardware pin, hardware connector".

As shown in Figure 5, HDA library should be developed further with the following elements in consideration of many core processors:

"Analog Digital Converters, Watchdog, RAM, ROM, GTM unit, Individual cores of many core processors, System peripheral bus and additional global memory elements".

FIGURE 5 Developed hardware library elements considering multicore processors.

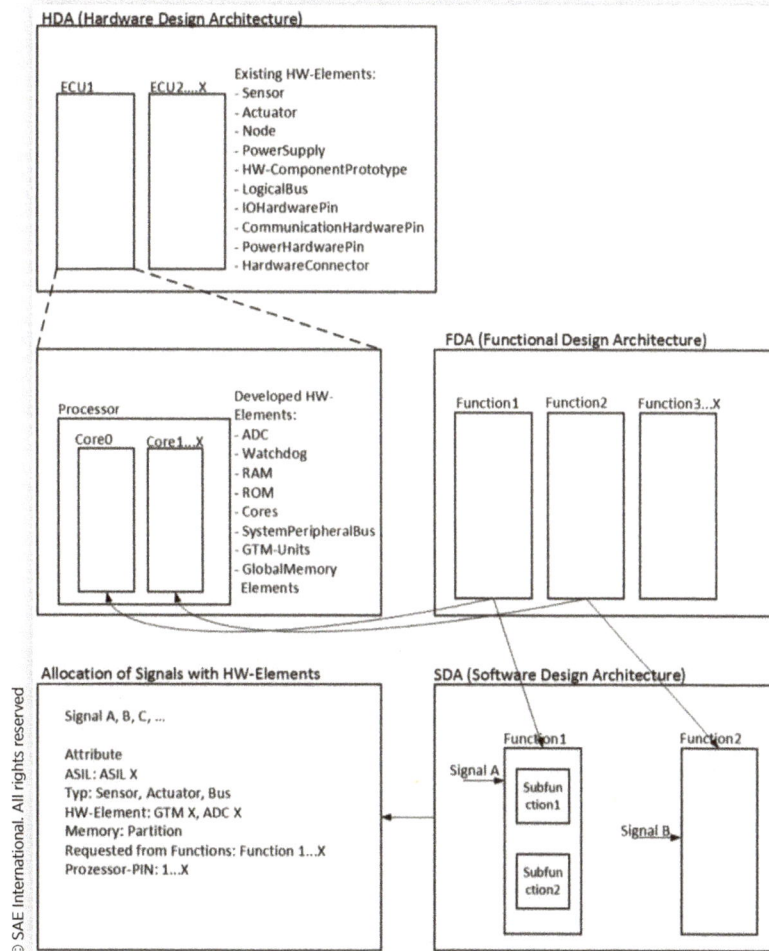

B. DESCRIPTION OF DFA-ANALYSIS AND SAFETY ANALYSIS

Figure 6 shows the basic concept of the DFA analysis. The relationship of the development steps and decomposition paths are modeled within EAST-ADL. There are two way to analyze the architecture. Top-down method enables to find the decomposition path from the model data. And the bottom-up method is used to provide the evidence that the sufficient independence between the architectural elements and signals exists.

As shown in Figure 6, the system modeling and safety modeling are merged together. It is necessary to extend the dependability modeling as safety goal definition with additional attributes, hardware design architecture (HDA) regarding multicore processor elements and software design architecture (SDA) within an HW-Elements allocation table for the signals. This allocation table contains the signals which are used for the safety related functions and the signal attributes as ASIL classification, type etc. And there are also additional information about the processor pin which belongs to which signal and the signal processing which analog digital converter (ADC) and timer modules (GTM, CCU6, GPT12, etc.) are used to create the signal. These extensions and information are

FIGURE 6   Model based DFA-Analysis.

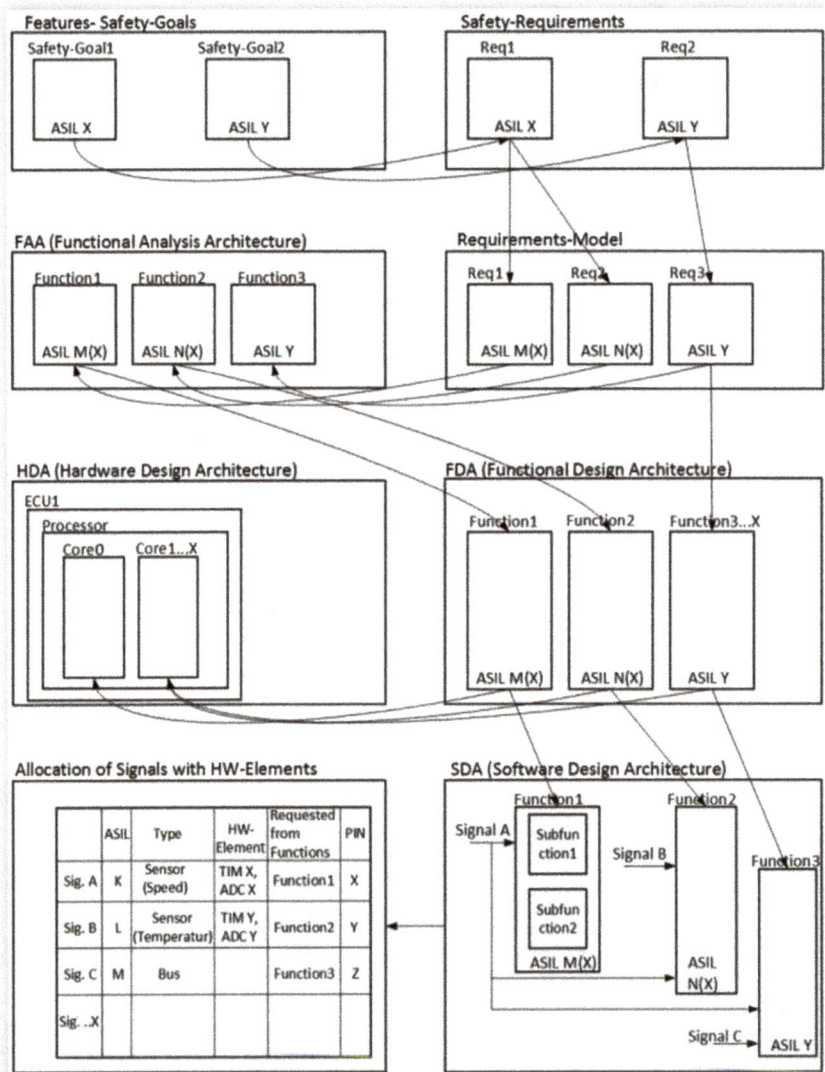

necessary in order to perform the DFA analysis automatically and in order to provide the evidence about sufficient independency between decomposed function parts.

This method enables the traceability over the general system model and shows the relationship between the system and safety architecture. Hence it is very easy to find out which safety goal and requirements are implemented in which software safety function and runs on which ECU and core. And the signals can be traceable also to the requirements, functions and HW-Elements.

The following DFA analysis rules are developed for the model check to analyze the safety concept compliant:

1. Relation Check:
   a. Safety-Goal1 -- > Req1 (Safety-Requirements) Safety-Goal2 -- > Req2 (Safety-Requirements)
   b. Req1 (Safety-Requirements) -- > Req1 (Requirements-Model) -- > Req2 (Requirements Model)
   Decomposition (one-to-many relationship)
2. ASIL Check:
   a. ASIL of Safety-Goal1 = ASIL of Req1 (Safety-Requirements)
   ASIL of Safety-Goal2 = ASIL of Req2 (Safety-Requirements)
   b. ASIL of Req1 (Safety-Requirements) = ASIL (in brackets) of Req1 and Req2 (Requirements-Model)
   ASIL of Req2 (Safety-Requirements) = ASIL of Req3 (Requirements-Model)
   c. ASIL of Function1 (FAA) = ASIL of Req1 (Requirements-Model)
   ASIL of Function2 (FAA) = ASIL of Req2 (Requirements-Model)
   ASIL of Function3 (FAA) = ASIL of Req3 (Requirements-Model)
   d. ASIL of Function1 (FAA) = ASIL of Function1 (FDA)
   ASIL of Function2 (FAA) = ASIL of Function2 (FDA)
   ASIL of Function3 (FAA) = ASIL of Function3 (FDA)
   e. ASIL of Function1 (FDA) = ASIL of Function1 (SDA)
   ASIL of Function2 (FDA) = ASIL of Function2 (SDA)
   ASIL of Function3 (FDA) = ASIL of Function3 (SDA)
3. Independency Check:
   a. Core (HDA) of Function1 (FDA) != Core (HDA) of Function 2 (FDA)
   b. Core (HDA) of Function3 (FDA) can be same as the core of other functions, because the Function3 is not part of decomposition.
4. Signal Check:
   a. ASIL of Signal A >= ASIL of Function1 (SDA)
   ASIL of Signal B >= ASIL of Function2 (SDA)
   ASIL of Signal C >= ASIL of Function3 (SDA)
   ASIL of Signal A >= ASIL of Function3 (SDA)
   b. TIM and ADC of Signal A != TIM and ADC of Signal B
   c. Input signals of Function1 != Input signals of Function 2
   d. Output signals of Function 1 != Input signals of Function 2

Figure 7 shows the developed DFA-Analysis-Tool which contains the rules to check the independency. The result of the analysis is also listed in the same tool.

FIGURE 7 DFA-Analysis-Tool and results.

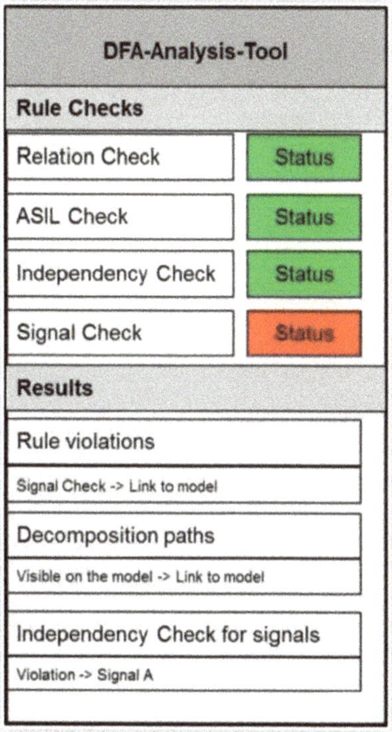

CHAPTER 3

C. USE-CASE AND REPORTS

Figure 7 shows additionally the DFA analysis results. In this case the independency of signals is violated by the decomposition path. Signal A in the Figure 8 is requested from two different functions which are the part of decomposition and therefore should be independent of each other.

The following example on Figure 8 shows the relationship between safety goals, safety requirements and safety functions. After the analysis, the decomposition path and additional corresponding elements are marked with different color to find the independency path easier. Another advantage of this representation is to recognize the decomposition paths. This protects against the violation of independency rules by providing a new signal. The signals are allocated with hardware elements such as TIMs and ADCs in order to analyze hardware independency.

FIGURE 8 Use-Case and reports.

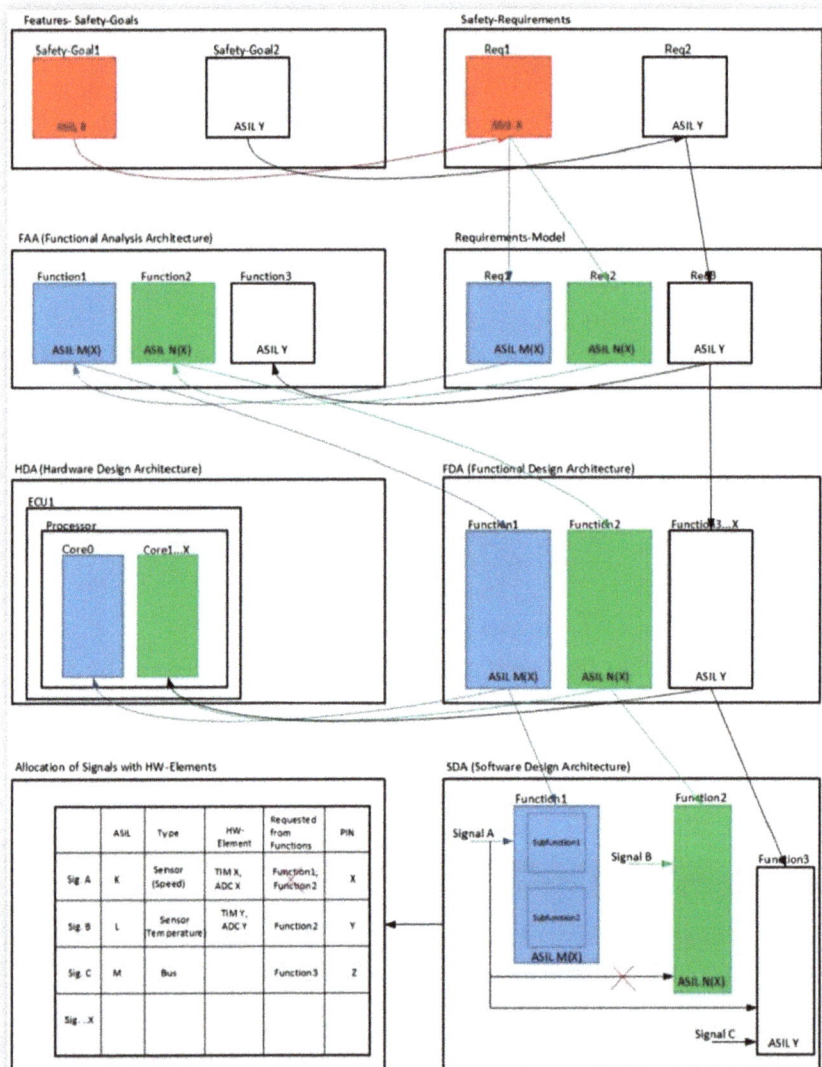

Conclusions

The presented approach and developed automated model based DFA analysis support the developers in creating safety architecture and providing evidence of sufficient independence between decomposed architectural elements.

The proposed model based DFA-Analysis enables to check automatically whether the evidence of sufficient independence between decomposition paths and signals exist. The automation of this step reduces the development effort because the decomposition path from system architecture to software and hardware architecture can be analyzed through model based approach. Traceability ensures that the additional used signal can be checked in early design stage before using it in a safety function. So it is very efficient to find out all the signals in the earlier development phase which are not allowed to be used because of decomposition violation.

The other important advantage of this method is the traceability of the different steps. Once user is familiar with the method, it is easy to get understanding of overall architecture. Therefor the approach enables to understand the system and safety architecture of an item very quickly. With the additional extensions of EAST-ADL regarding multicore processor related elements, it is now possible to create the architectural design of multicore processors. It also allows the modeling of a multicore processor with its hardware elements and software safety architecture which are necessary to prove hardware and software independency. The extensions and developed scripts make it possible to gain sufficient transparency and traceability for the safety arguments and to support the whole safety process for multicore processors in a single solution even in hardware and software development. Within this approach, requirement modeling, dependency modeling and hardware modeling of EAST-ADL are extended in order to model the safety-critical functions with multicore processors. The user can thus find out easily which requirements are implemented for which functions and are located on which core.

As future work this approach can be extended further in such a way that the safety mechanisms of detected common causes are considered by DFA-Analysis. So it can be check automatically whether the identified common causes lead to a safe state.

Contact Information

Bülent Sari
ZF Friedrichshafen
AG Graf-von-Soden-Platz 1
88046 Friedrichshafen
Germany
Buelent.sari@zf.com

Acknowledgments

The realization of DFA analysis is not limited to any particular modeling language or tool.

References

1. Charette, R.N., "This Car Runs on Code," http://www.realprogrammer.com/interesting_things/IEEE%20SpectrumThisCarRunsOnCode.pdf, February 2009.

2. PWC DEUTSCHLAND, "Autoindustrie treibt Chipnachfrage an," http://www.pwc.de/de/automobilindustrie/autoindustrie-treibtchipnachfrage-an.html, Version, 2013.

3. AK-L_Orientation-list-V1.2_2010-11-25_DE (AA-I3/AK 16 -Functional Safety).

4. Hans, B., Henrik, L.; Frank, H.; Yiannis, P. et al., "EAST-ADL - An Architecture Description Language for Automotive Software-Intensive Systems - White Paper Version 2.1.12," http://www.maenad.eu/public/conceptpresentations/EAST-ADL_WhitePaper_M2.1.12.pdf, 2013.

5. Cuenot, P., Chen, D., Gerard, S., Lonn, H. et al., "Managing Complexity of Automotive Electronics Using the EAST-ADL," *Engineering Complex Computer Systems, 2007. 12th IEEE International Conference on*, 2007.

6. ATESST2: ATESST2, http://www.atesst.org, 2010.

7. MAENAD: MAENAD, http://www.maenad.eu/, 2014.

8. SAFE: SAFE, http://www.safe-project.eu/, 2014.

9. AUTOSAR: AUTOSAR, http://www.autosar.org/, 2016.

10. HIP-HOPS: HiP-HOPS, "Automated Fault Tree, FMEA and Optimisation Tool," http://www.hiphops.eu/index.php/the-manual, 2013.

11. ALTARICA: ALTARICA, https://altarica.labri.fr/forge/.

12. Sari, B. and Reuss, H.C., "A Model-Driven Approach for the Development of Safety-Critical Functions Using Modified Architecture Description Language (ADL)," *Electrical Systems for Aircraft, Railway, Ship propulsion and Road Vehicles & International Transportation Electrification Conference (ESARS-ITEC)*, 2016.

13. ISO 26262 - Part 9, "Requirements Decomposition with Respect to ASIL Tailoring," 2011.

4

An Analysis of ISO 26262: Machine Learning and Safety in Automotive Software

Rick Salay, Rodrigo Queiroz, and Krzysztof Czarnecki
University of Waterloo

Machine learning (ML) plays an ever-increasing role in advanced automotive functionality for driver assistance and autonomous operation; however, its adequacy from the perspective of safety certification remains controversial. In this paper, we analyze the impacts that the use of ML within software has on the ISO 26262 safety lifecycle and ask what could be done to address them. We then provide a set of recommendations on how to adapt the standard to better accommodate ML.

Introduction

The use of machine learning (ML) is on the rise in many sectors of software development, and automotive software development is no different. In particular, Advanced Driver Assistance Systems (ADAS) and Automated Driving Systems (ADS) are two areas where ML plays a significant role [1, 2]. In automotive development, safety is a critical objective, and the emergence of standards such as ISO 26262 [3] has helped focus industry practices to address safety in a systematic and consistent way. Unfortunately, ISO 26262 was not designed to accommodate technologies such as ML, and this has created a tension between the need to innovate and the need to improve safety.

In response to this issue, research has been active in several areas. Recently, the safety of ML approaches in general have been analyzed both from theoretical [4] and pragmatic perspectives [5]. However, most research is specifically about neural networks (NN). Work on supporting the verification & validation (V&V) of NNs emerged in the 1990's with a focus on making their internal structure easier to assess by extracting representations that are more understandable [6]. General V&V methodologies for NNs have also been proposed [7, 8]. More recently, with the popularity of deep neural networks (DNN), verification research has included more diverse topics such as generating explanations of DNN predictions [9], improving the stability of classification [10] and property checking of DNNs [11].

The recent surge of interest in ADSs has also been driving research in certification. Despite their challenges, NNs are already used in high assurance systems (see [12] for a survey), and safety certification of NNs has received some attention. Pullum et al. [13] give detailed guidance on V&V as well as other aspects of safety assessment such as *hazard analysis* with a focus on adaptive systems in the aerospace domain. Bedford et al. [14] define general requirements for addressing NNs in any safety standard. Kurd et al. [15] have established criteria for NNs to use in a safety case.

The recent surge of interest in ADSs has also been driving research in certification. Koopman and Wagner [2] identify some of the key challenges to certification, including ML. Martin et al. [16] analyze the adequacy of ISO 26262 for an ADS but focuses on the impact of the increased complexity it creates rather than specifically the use of ML. Spanfelner et al. [1] assess ISO 26262 from the perspective of driver assistance systems. Finally, Burton et al. [17] explore the kind of safety case that is required for an ADS that uses ML components.

The contribution of the current paper is complementary to the above research. We analyze the impact that the use of ML-based software has on various parts of ISO 26262. Specifically, we consider its impact in the areas of hazard analysis and in the phases of the software development process. In all, we identify five distinct problems that the use of ML creates and make recommendations on steps toward addressing these problems both through changes to the standard and through additional research.

The remainder of the paper is structured as follows. In the next section we give the required background on ISO 26262 and ML. Following this is the analysis of the ISO 26262 safety lifecycle with five subsections describing each impacted area and the corresponding recommendations. Finally, we summarize and give concluding remarks.

Background

ISO 26262

ISO 26262 is a standard that regulates functional safety of road vehicles. It recommends the use of a Hazard Analysis and Risk Assessment (HARA) method to identify hazardous events in the system and to specify safety goals that mitigate the hazards. The standard has 10 parts, but we focus on Part 6: "product development at the software level". The standard follows the well-known V model for engineering shown in Figure 1.

An Automotive Safety Integrity Level (ASIL) refers to a risk classification scheme defined in ISO 26262 for an item (e.g., subsystem) in an automotive system. The ASIL represents the degree of rigor required (e.g., testing techniques, types of documentation required, etc.) to reduce the risk of the item, where ASIL D represents the highest and ASIL A the lowest risk. If an element is assigned QM (Quality Management), it does not require safety management. The ASIL assessed for a given hazard is first assigned to the safety goal set to address the hazard and is then inherited by the safety requirements derived from that goal.

FIGURE 1   ISO 26262 part 6 - Product development at the software level.

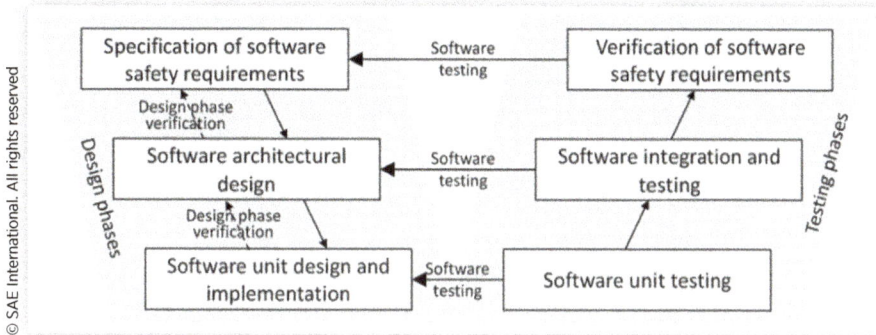

FIGURE 2   ISO 26262 Part 6 - Mechanisms for error handling at the software architectural level.

Methods		ASIL			
		A	B	C	D
1a	Static recovery mechanism	+	+	+	+
1b	Graceful degradation	+	+	++	++
1c	Independent parallel redundancy	o	o	+	++
1d	Correcting codes for data	+	+	+	+

Part 6 of the standard specifies the compliance requirements for software development. For example, Figure 2 shows the error handling mechanisms recommended for use as part of the architectural design. The degree of recommendation for a method depends on the ASIL and is categorized as follows: ++ indicates that the method is highly recommended for the ASIL; + indicates that the method is recommended for the ASIL; and o indicates that the method has no recommendation for or against its usage for the ASIL. For example, *Graceful Degradation* (1b) is the only highly recommended mechanism for an ASIL C item, while an ASIL D item would also require *Independent Parallel Redundancy* (1c).

Machine Learning

In this paper, we are concerned with software implementation using ML. We call a *programmed component* to be one that is implemented using a programming language, regardless of whether the programming was done manually or automatically (e.g., via code generation). In contrast, an *ML component* is one that is a trained model using a supervised, unsupervised or reinforcement learning (RL) approach.

An ML component can be trained offline during system development or online in a running system. For ML components in automotive systems, we assume that online learning is limited to non safety-critical functionality. For example, an ML component could be trained online to learn a driver's infotainment preferences. The key weakness of online learning with respect to functional safety assurance is that a safety argument cannot be made and assessed ahead of time. Thus, for the applications of ML discussed in this paper, we assume training is done offline.

There are several characteristics of ML that can impact safety or safety assessment.

NON-TRANSPARENCY

All types of ML models contain knowledge in an encoded form, but this encoding is more *transparent* - i.e., easier to interpret by humans in some types than others. Bayesian Networks are transparent since the nodes are random variables and can represent human-defined concepts. For example, a Bayesian Network for weather prediction may have nodes such as "precipitation type", "temperature", etc. In contrast, NN models are considered non-transparent and significant research effort has been devoted to making them more transparent (e.g., [6, 9]). Increasing ML model expressive power is typically at the expense of transparency but some research efforts focus on mitigating this [18]. Non-transparency is an obstacle to safety assurance because it is more difficult for an assessor to develop confidence that the model is operating as intended.

ERROR RATE

An ML model typically does not operate perfectly and exhibits some error rate. Thus, "correctness" of an ML component, even with respect to test data, is seldom achieved and it must be assumed that it will periodically fail. Furthermore, although an estimate of the true error rate is an output of the ML development process, there is only a statistical guarantee about the reliability of this estimate. Finally, even if the estimate of the true error rate was accurate, it may not reflect the error rate the system actually experiences while in operation after a finite set of inputs because the true error is based on an infinite set of samples [4]. These characteristics must be considered when designing safe system using ML components.

TRAINING-BASED

Supervised and unsupervised learning based ML models are trained using a subset of possible inputs that could be encountered operationally. Thus, the training set is necessarily incomplete and there is no guarantee that it is even representative of the space of possible inputs. In addition, learning may overfit a model by capturing details incidental to the training set or training environment rather than general to all possible inputs in the operational environment [17]. RL suffers from similar limitations since it typically explores only a subset of possible behaviours during training. The uncertainty that this creates about how an ML component will behave is a threat to safety. Another factor is that, even if the training set is representative, it may under-represent the safety-critical cases because these are often rarer in the input space [4]. Finally, over time, the underlying distribution of inputs in the operational environment may drift from that of the training set, degrading safety [17].

INSTABILITY

More powerful ML models (e.g., DNN) are typically trained using local optimization algorithms, and there can be multiple optima. Thus, even when the training set remains the same, the training process may produce a different result. However, changing the training set also may change the optima. In general, different optima may be far apart structurally, even if they are similar behaviourally. This characteristic makes it difficult to debug models or reuse parts of previous safety assessments.

Analysis of ISO 26262

In this section, we detail our analysis of ML impacts on ISO 26262. Since ML-based software is a specialized type of software, we classify an area of the standard as *impacted* when it is relevant to software and the treatment of ML-based software should differ from the existing treatment of software by the standard. Applying this criterion to the ten parts of the standard resulted in identifying five areas of impact in two parts: the hazard analysis from the concept phase (Part 3) and the software development phase (Part 6). We describe the five areas of impact with corresponding recommendations in the following subsections. Where it is relevant, we also indicate the levels of autonomy (i.e., SAE J3016 [19] levels 0-5) to which the impact applies.

Identifying Hazards

ISO 26262 defines a hazard as "a potential source of harm caused by malfunctioning behaviour of the item where harm is physical injury or damage to the health of persons" [3, Part 1]. The use of ML can create new types of hazards. One type of such hazard applicable at automation levels 1-3 is caused by the human operator becoming complacent because they think the automated driver assistance (often using ML) is smarter than it actually is [20]. For example, the driver stops monitoring steering in an automated steering function. On one level, this can be viewed as a case of "reasonably forseeable misuse" by the operator, and such misuse is identified in ISO 26262 as requiring mitigation [3, Part 3]. However, this approach may be too simplistic. As ML creates opportunities for increasingly sophisticated driver assistance, the role of the human operator becomes increasingly critical to correct for malfunctions. But increasing automation can create behavioural changes in the operator, reducing their skill level and limiting their ability to respond when needed [21]. Such behavioural impacts can negatively impact safety even though there is no system malfunction or misuse.

Other new types of hazards are due to the unique ways an ML component can fail at higher automation levels (i.e., 4 and 5). For RL, faults in the reward function can cause surprising failures. An RL-based component may negatively affect the environment in order to achieve its goal [5]. For example, an ADS may break laws in order to reach a destination faster. Another possibility is that the RL component *games* the reward function [5]. For example, the ADS figures out that it can avoid getting penalized for driving too close to other cars by exploiting certain sensor vulnerabilities so that it cannot "see" how close it is getting. Although hazards such as these may be unique to ML components, they can be traced to faults, and thus they fit within the existing guidelines of ISO 26262.

RECOMMENDATIONS FOR ISO 26262

The definition of hazard should be broadened to include harm potentially caused by complex behavioural interactions between humans and the vehicle that are not due to a system malfunction. The standard itself takes note that the current definition is "restricted to the scope of ISO 26262; a more general definition is potential source of harm"[3, Part 1]. The definition and methods for identifying such hazards should be informed by the research specifically on behavioural impacts of ADAS [22] as well as human-robot interaction (HRI) [23] more broadly. For example, van den Brule et al. [24] study how a robot's behavioural style can affect the trust of humans interacting with it.

Faults and Failure Modes

ISO 26262 mandates the use of analyses such as Fault Mode Effects Analysis (FMEA) to identify how faults lead to failures that may cause harm (i.e., are hazards). We can ask whether there are types of faults and failures that are unique to ML and not found in programmed software. Specific fault types and failure modes have been catalogued for NNs (e.g., [13, 15]). Some of these are just "apparent" ML specific faults. For example, a neuron that randomly changes its connection in an operational NN is not really about neurons but rather a conventional fault that can occur in the software on which the NN runs. Others are distinctly ML-specific such as faults in the network topology and learning method that lead to poor generalization (e.g., insufficient connectivity between layers, too high a learning rate, etc.) or faults in the training set. These include inadequate representativeness of the operational environment by the training set, insufficient coverage of rare cases and lack of handling for distributional shift.

Although ML faults have some unique characteristics, this cannot be said about failure modes. All that ML faults can do is to increase the error rate of the deployed component, and thus cause one particular type of failure - an incorrect output for some input. But since most software failures take the form of incorrect output for a given input, we may conclude that there is nothing different about the failure analysis of an ML component as compared to a programmed component, and existing ISO 26262 recommendations apply.

RECOMMENDATIONS FOR ISO 26262

The distinctive types of ML faults create the opportunity to develop focused tools and techniques to help find faults independently of the domain for which the ML model is being trained. For example, Chakarov et al. [25] describe a technique for debugging misclassifications due to bad training in data, while Nushi et al. [26] propose an approach for troubleshooting faults due to complex interactions between linked ML components. As these techniques mature, ISO 26262 should be amended to require the use of such techniques for ML components.

When the functionality is complex and the ASIL is high (e.g., at higher automation levels), it is unlikely that the error rate can ever be brought to an acceptably low level only through increasing or improving the training set due to the "curse of dimensionality". Specialized architectural techniques should be required to help mitigate the effects of ML faults and failures. Ensemble methods [27] such as bagging [28] and boosting [29] are mature "fault tolerance" techniques used with ML classifiers that reduce the error rate by fusing the result of multiple weaker classifiers to produce a stronger classifier. The simplex architecture [30] that uses a conservative but verifiably safe controller as a fall-back from a more advanced but unverifiable controller has been proposed as a way to make ML components safe [31]. Other similar "safety envelope" approaches have also been proposed for reinforcement learning. [32, 33].

Specification and Verification

Spanfelner et al. [1] point out that there is an assumption in ISO 26262, given by the left side of the V model (Figure 1), that component behaviour is fully specified and each refinement can be verified with respect to its specification. Note that this assumption is also made in other safety-critical domains such as aerospace [34]. This is important to ensure that a safety argument can trace the behaviour of the implementation to its design, safety requirements and ultimately, to the hazards that are mitigated.

This assumption is violated when a training set is used in place of a specification since such a set is necessarily incomplete, and it is not clear how to create assurance that the corresponding hazards are always mitigated. Thus, an ML component violates the assumption. Furthermore, the training process is not a verification process since the trained model will be "correct by construction" with respect to the training set, up to the limits of the model and the learning algorithm.

A deeper issue, discussed by Spanfelner et al. [1], is that many kinds of advanced functionality needed for higher automation levels require perception of the environment, and this functionality may be *inherently unspecifiable*. For example, what is the specification for recognizing a pedestrian? We might observe that since a vehicle must move around in a human world, advanced functionality must involve perception of *human categories* (e.g., pedestrians). There is evidence that such categories can only partially be specified using rules (e.g., necessary and sufficient conditions) and also need examples [35]. This suggests that ML-based approaches are necessary for implementing this type of functionality.

RECOMMENDATIONS FOR ISO 26262

The approach required for high ASIL component implementation should be based on the specifiability of the functionality being implemented. For functionality that is fully specifiable, programming must be required. For functionality that admits no complete specification (e.g., perception), ML-based approaches should be allowable, and the complete specification requirement must be relaxed. However, even here, the use of specification in a more limited capacity should still be required where possible. The higher-level safety requirements for an ML component allocated by the architectural level (i.e., before refinement at the unit level) can be specified with completeness and traced to hazards. For example, a component may have the requirement "identify pedestrians" in order that the ADS could avoid harming them.

At the component unit design level, the requirements on sample data can be specified in order to ensure that appropriate training, validation and testing sets are obtained. Subsequently, the data gathered can be verified with respect to this specification. Techniques from black-box software testing such as input domain partitioning may be helpful here [36]. For example, the input domain of a pedestrian classifier could be partitioned by age category, pose (e.g., standing, leaning, etc.), clothing type, etc. The data requirements can specify the relative numbers of samples that should come from each partition to ensure coverage and representativeness of the sample data.

Finally, although complete behavioral specifications are not possible, partial specifications may still be. For example, if a pedestrian must be less than 9 feet tall, then this property can be used to filter out false positives. Such properties can be incorporated into the training process or checked on models after training (e.g., [11]). Some of these recommendations may be addressed in a forth-coming OMG standard relating to sensor and perception issues [37].

Level of ML Usage

Figure 1 identifies an architectural level and a unit (i.e., component) level of implementation. ISO 26262 defines a software architecture as consisting of components and their interactions in a hierarchical structure [3, Part 6]. This component decomposition is important for safety because it allows for easier comprehension of a complex system by human assessors and it permits the use of compositional formal analysis techniques.

ML could be used to implement an entire software system, including its architecture, using an *end-to-end* approach. For example, Bojarski et al. [38] train a DNN to make the appropriate steering commands directly from raw sensor data, sidestepping typical ADS architectural components such as lane detection, path planning, etc.

Here, we may assume that the unit level, in the conventional sense of a distinct component that can be developed independently of the architecture, *no longer exists*. This is the case, even if it is possible to extract and interpret the structure of the trained model as consisting of units with distinct functions, since this structure is emergent in the training process and unstable. If the model is re-trained with a slightly different training set, this structure can change arbitrarily. Note that a DNN does have an

CHAPTER 4

architecture in a different sense - the set of layers and their connections. However, since it is the training that actually "implements" the required functionality, this architecture is more of an generic execution layer. Thus, an end-to-end approach deeply challenges the assumptions underlying ISO 26262.

Another challenge with an end-to-end approach is that, in some cases, the size of the training set needs to be exponentially larger than when a programmed architecture is used [39]. This puts additional strain on the already challenging problem of obtaining an adequate training set for safety-critical contexts.

Finally, note that issues with an end-to-end approach can also apply when ML is used at the component level, if components are too complex. For example, at one extreme, the architecture can consist of a single component. ISO 26262 specifically guards against this pitfall by mandating the use of modularity principles such as restricting the size of components and maximizing the cohesion within a component. However, the lack of transparency of ML components can hamper the ability to assess component complexity and therefore, to apply these principles. Fortunately, improving ML transparency is an active research area (e.g., [9, 6]).

RECOMMENDATIONS FOR ISO 26262

Although using an end-to-end approach has shown some recent successes with autonomous driving (e.g., [38]), it is incompatible with the assumptions about stable hierarchical architectures of components. This limits the use of most techniques for system safety and we therefore recommend that ISO 26262 only allow the use of ML at the component level.

Required Software Techniques

Part 6 of ISO 26262 deals with product development at the software level and specifies 75 software development techniques, such as shown in Figure 2, that are used in various phases of the development process in the V model (Figure 1). Of these, 34 apply at the unit level, and the remaining at the architectural level. We performed an assessment of the software techniques to determine their applicability to ML components*. Based on our recommendation above on the level of ML usage, we assumed that ML was only used at the unit level and programming is used at the architecture level to connect components.

The charts in Figure 3 show the results of the assessment for the techniques dealing with the unit level. We classified each technique into one of three categories based on the level of applicability to ML. Category Ok means the technique is directly applicable without modification. Most of these cases are due to the fact that they are black box techniques (e.g., *analysis of boundary values*, *error guessing*, etc.) and thus, the method of component implementation is irrelevant. However, some white box techniques such as *fault injection* also apply. For example, faults can be injected into an NN by breaking links or randomly changing weights (e.g., [40]). Category Adapt says that the technique can be used for an ML component if it is adapted in some way. For example, the technique *walk-through* cannot be used directly with an NN due to the non-transparency characteristic. Finally, category N/A indicates that the technique is fundamentally code-oriented and does not apply to an ML component. For example, *no multiple use of variable names* is meaningful for a program but has no corresponding notion in an ML model.

---

* The data is available at https://github.com/rsalay/safetyml

FIGURE 3 Percentage of unit-level software techniques applicable to ML components: (a) values averaged across the four ASILs with standard deviation shown in parentheses; (b) values for each ASIL when only highly recommended techniques are considered.

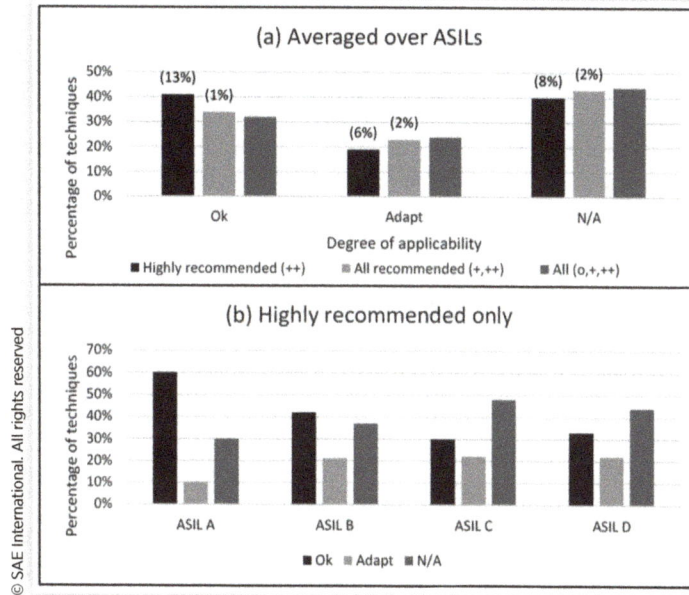

The results in Chart (a) are grouped by the degree to which the techniques are recommended. Recall from background section that each technique is marked as highly recommended (++), recommended (+) or no recommendation (o) depending on the ASIL level. The bars in each category show the percentage of techniques that apply when considering all techniques (0,+,++), only the recommended techniques (+, ++), and only the highly recommended techniques (++). Since the degree of recommendation varies by ASIL, each percentage is an average value over all four ASILs with the standard deviation in parentheses. Note that the standard deviation is 0 for the "all" group since every technique is present for each ASIL. Because of the high standard deviation for the highly recommended group, we have included Chart (b) which gives the actual data for each ASIL in this group.

Chart (a) shows that a significant part of the standard is still directly applicable (category Ok) and there is an emphasis on highly recommended techniques. However, the standard deviation is high and Chart (b) shows that most of these highly recommended techniques apply to the lower ASIL values - i.e. they are less rel evant from a safety critical perspective. Chart (a) also shows that about 40% of the techniques do not apply at all (category N/A) regardless of the degree of recommendation. In general, techniques in the software part of the standard are clearly biased toward imperative programming languages (e.g., C, Java, etc.) [34]. In addition to precluding ML components, this bias makes it difficult to accept implementations in other mature programming paradigms such as functional programming, logic programming, etc.

RECOMMENDATIONS FOR ISO 26262

One approach to addressing the gap in applicable techniques as well as the imperative language bias without compromising safety may be to specify the requirements for techniques based on their *intent* and maturity rather than on their specific details. For example, the intent of the *no multiple use of variable names* technique is to reduce the possibility for confusion that may prevent the detection of bugs. This helps humans understand the implementation better and increase their confidence in its correctness and safety. Thus, the standard can require the use of "accepted clarity increasing" techniques instead of the specific techniques.

Summary and Conclusion

Machine learning is increasingly seen as an effective software implementation technique for delivering advanced functionality; however, how to assure safety when ML is used in safety critical systems is still an open question. The ISO 26262 standard for functional safety of road vehicles provides a comprehensive set of requirements for assuring safety but does not address the unique characteristics of ML-based software. In this paper, we make a step towards addressing this gap by analyzing the places where ML can impact the standard and providing recommendations on how to accommodate this impact. Our results and recommendations are summarized as follows.

Identifying Hazards

The use of ML can create new types of hazards that are not due to the malfunctioning of a component. In particular, the complex behavioural interactions possible between humans and advanced functionality implemented by ML can create hazardous situations that should be mitigated within the system design. We recommend that ISO 26262 expands their definition of hazard to address these kinds of situations.

Fault and Failure Modes

ML components have a development lifecycle that is different from other types of software. Analyzing the stages in the lifecycle reveals distinct types of faults they may have. We recommend that ISO 26262 be extended to explicitly address the ML lifecycle and require the use of fault detection tools and techniques that are customized to this lifecycle.

Specification and Verification

Because ML components are trained from inherently incomplete data sets, they violate the assumption in V model-based processes that component functionality must be fully specified and that refinements are verifiable. Furthermore, it is possible that certain types of advanced functionality (e.g., requiring perception) for which ML is well suited are unspecifiable in principle. As a result, ML components are designed with the knowledge that they have an error rate and that they will periodically fail. Rather than disqualifying this class of functionality, we recommend that ISO 26262 provide different safety requirements depending on whether the functionality is specifiable.

The Level of ML Usage

ML could be used broadly at the architectural level with a system by using an end-to-end approach or remain limited to use at the component level. The end-to-end approach challenges the assumption that a complex system is modeled as a stable hierarchical decomposition of components each with their own function. This limits the use of most techniques for system safety and we therefore recommend that ISO 26262 only allow the use of ML at the component level.

Required Software Techniques

ISO 26262 mandates the use of many specific techniques for various stages of the software development lifecycle. Our analysis shows that while some of these remain applicable to ML components and others could readily be adapted, many remain that are specifically biased toward the assumption that code is implemented using an imperative programming language. In order to remove this bias, we recommend that the requirements be expressed in terms of the intent and maturity of the techniques rather than their specific details.

Acknowledgment

The authors would like to thank Atrisha Sarkar, Michael Smart, Michal Antkiewicz, Marsha Chechik, Sahar Kokaly and Ramy Shahin for their insightful comments.

References

1. Spanfelner, B., Richter, D., Ebel, S., Wilhelm, U., Branz, W., and Patz, C., "Challenges in Applying the ISO 26262 for Driver Assistance Systems," *Tagung Fahrerassistenz, München* 15(16), 2012.

2. Koopman, P. and Wagner, M., "Challenges in Autonomous Vehicle Testing and Validation," *SAE Int. J. Trans. Safety* 4(1):15-24, 2016, doi:10.4271/2016-01-0128.

3. International Organization for Standardization, "ISO 26262: Road Vehicles - Functional Safety," 2011.

4. Varshney, K.R., "Engineering Safety in Machine Learning," arXiv preprint arXiv:1601.04126, 2016.

5. Amodei, D., Olah, C., Steinhardt, J., Christiano, P., Schulman, J., and Mané, D., "Concrete Problems in AI Safety," arXiv preprint arXiv:1606.06565, 2016.

6. Tickle, A.B., Andrews, R., Golea, M., and Diederich, J., "The Truth will Come to Light: Directions and Challenges in Extracting the Knowledge Embedded within Trained Artificial Neural Networks," *IEEE Transactions on Neural Networks* 9, no. 6 (1998): 1057-1068, doi:10.1109/72.728352.

7. Peterson, G.E., "Foundation for Neural Network Verification and Validation," *Optical Engineering and Photonics in Aerospace Sensing*, International Society for Optics and Photonics, 1993, 196-207.

8. Rodvold, D.M., "A Software Development Process Model for Artificial Neural Networks in Critical Applications," *International Joint Conference on Neural Networks*, 5 (1999): 3317-3322, doi:10.1109/IJCNN.1999.836192.

9. Hendricks, L.A., Akata, Z., Rohrbach, M., Donahue, J., Schiele, B., and Darrell, T., "Generating Visual Explanations," *European Conference on Computer Vision*, Spring, 2016, 3-19, doi:10.1007/978-3-319-46493-01.

CHAPTER 4

10. Huang, X., Kwiatkowska, M., Wang, S., and Wu, M., "Safety Verification of Deep Neural Networks," arXiv preprint arXiv:1610.06940, 2016.

11. Katz, G., Barrett, C., Dill, D., Julian, K., and Kochenderfer, M., "Reluplex: An Efficient SMT Solver for Verifying Deep Neural Networks," arXiv preprint arXiv:1702.01135, 2017.

12. Schumann, J., Gupta, P., and Liu, Y., "Application of Neural Networks in High Assurance Systems: A Survey," *Applications of Neural Networks in High Assurance Systems*, Spring, 2010, 1-19, doi:10.1007/978-3-642-10690-3.

13. Pullum, L. L., Taylor, B. J., and Darrah, M. A., *Guidance for the Verification and Validation of Neural Networks*, vol. 11 (John Wiley & Sons, 2007), doi:10.1002/9781119134671.

14. Bedford, D., Morgan, G., and Austin, J., "Requirements for a Standard Certifying the Use of Artificial Neural Networks in Safety Critical Applications," *Proceedings of the International Conference on Artificial Neural Networks*, 1996.

15. Kurd, Z., Kelly, T., and Austin, J., "Developing Artificial Neural Networks for Safety Critical Systems," *Neural Computing and Applications* 16, no. 1 (2007): 11-19, doi:10.1007/s00521-006-0039-9.

16. Martin, H., Tschabuschnig, K., Bridal, O., and Watzenig, D., "Functional Safety of Automated Driving Systems: Does ISO 26262 Meet the Challenges?," *Automated Driving*, Spring, 2017, 387-416, doi:10.1007/978-3-319-31895-0_16.

17. Burton, S., Gauerhof, L., and Heinzemann, C., "Making the Case for Safety of Machine Learning in Highly Automated Driving," *International Conference on Computer Safety, Reliability, and Security*, Spring, 2017, 5-16, doi:10.1007/978-3-319-66284-8_1.

18. Henzel, M., Winner, H., and Lattke, B., "Herausforderungen in der Absicherung von Fahrerassistenzsystemen bei der Benutzung maschinell gelernter und lernender Algorithmen," *Proceedings of 11th Workshop Fahrerassistenzsysteme Und Automatisiertes Fahren (FAS)*, 2017, 136-148.

19. SAE International Surface Vehicle Recommended Practice, "Taxonomy and Definitions for Terms Related to On-Road Motor Vehicle Automated Driving Systems," SAE Standard J3016, 2017.

20. Parasuraman, R. and Riley, V., "Humans and Automation: Use, Misuse, Disuse, Abuse," *Human Factors: The Journal of the Human Factors and Ergonomics Society* 39, no. 2 (1997): 230-253, doi:10.1518/001872097778543886.

21. Brookhuis, K. A., De Waard, D., and Janssen, W. H., "Behavioural Impacts of Advanced Driver Assistance Systems-An Overview, *EJTIR*, 1, no. 3 (2001): 245-253.

22. Sullivan, J.M., "Flannagan," Pradhan, A.K. and Bao, S., *Literature Review of Behavioral Adaptations to Advanced Driver Assistance Systems*. AAA Foundation for Traffic Safety, 2016.

23. Goodrich, M.A. and Schultz, A.C., "Human-Robot Interaction: A Survey," *Foundations and Trends in Human-Computer Interaction* 1, no. 3 (2007): 203-275, doi:10.1561/1100000005.

24. van den Brule, R., Dotsch, R., Bijlstra, G., Wigboldus, D. H., and Haselager, P., "Do Robot Performance and Behavioral Style Affect Human Trust?," *International Journal of Social Robotics*, 6, no. 4 (2014): 519-531, doi:10.1007/s12369-014-0231-5.

25. Chakarov, A., Nori, A., Rajamani, S., Sen, S., and Vijaykeerthy, D., "Debugging Machine Learning Tasks," arXiv preprint arXiv 1603:07292, 2016.

26. Nushi, B., Kamar, E., Horvitz, E., and Kossmann, D.," "On Human Intellect and Machine Failures: Troubleshooting Integrative Machine Learning Systems," arXiv preprint arXiv 1611:08309, 2016.

27. Ponti, M.P., "Combining Classifiers: From the Creation of Ensembles to the Decision Fusion," *24th SIBGRAPI Conference on Graphics, Patterns and Images Tutorials (SIBGRAPI-T)*, IEEE, 2011, 1-10, doi:10.1109/SIBGRAPIT.2011.9.

28. Breiman, L., "Bagging Predictors," *Machine Learning* 24, no. 2 (1996): 123-140, doi:10.1007/BF00058655.

29. Freund, Y., Schapire, R.E. et al., "Experiments with a New Boosting Algorithm," *ICML 96* (1996): 148-156.

30. Sha, L., "Using Simplicity to Control Complexity," *IEEE Software* 18, no. 4 (2001): 20-28.

31. Phan, D., Yang, J., Clark, M., Grosu, R. et al., "A Component-Based Simplex Architecture for High-Assurance Cyber-Physical Systems," arXiv preprint arXiv 1704:04759, 2017.

32. Perkins, T.J. and Barto, A.G., "Lyapunov Design for Safe Reinforcement Learning," *Journal of Machine Learning Research* 3 (2002): 803-832.

33. Fisac, J.F., Akametalu, A.K., Zeilinger, M.N., Kaynama, S. et al., "A General Safety Framework for Learning-Based Control in Uncertain Robotic Systems," arXiv preprint arXiv 1705:01292, 2017.

34. Bhattacharyya, S., Cofer, D., Musliner, D., Mueller, J., and Engstrom, E., "Certification Considerations for Adaptive Systems," *Unmanned Aircraft Systems (ICUAS), 2015 International Conference on*, IEEE, 2015, 270-279, doi:10.1109/ICUAS.2015.7152300.

35. Rouder, J.N. and Ratcliff, R., "Comparing Exemplar and Rule-Based Theories of Categorization," *Current Directions in Psychological Science* 15, no. 1 (2006): 9-13, doi:10.1111/j.0963-7214.2006.00397.x.

36. Ammann, P. and Offutt, J., "*Introduction to Software Testing*," Cambridge University Press, 2016, doi:10.1017/CBO9780511809163.

37. International Organization for Standardization, "ISO/AWI PAS 21448: Road Vehicles - Safety of the Intended Functionality," under development.

38. Bojarski, M., Del Testa, D., Dworakowski, D., Firner, B., Zhang, J. et al., "End to End Learning for Self-Driving Cars," arXiv preprint arXiv 1604:07316, 2016.

39. Shalev-Shwartz, S. and Shashua, A., "On the Sample Complexity of End-to-End Training vs. Semantic Abstraction Training," arXiv preprint arXiv 1604:06915, 2016.

40. Takanami, I., Sato, M., and Yang, Y.P., "A Fault-Value Injection Approach for Multiple-Weight-Fault Tolerance of MNNs," *Proceedings of the IEEE-INNS-ENNS International Joint Conference on Neural Networks* 3 (2000): 515-520, doi:10.1109/IJCNN.2000.861360.

Hazard Analysis and Risk Assessment beyond ISO 26262: Management of Complexity via Restructuring of Risk-Generating Process

Oleg Lurie and Joseph Miller
ZF - TRW

The automotive world is getting ready to embrace the automated driving (AD), while advanced driver assistance systems (ADAS) increase their authority in the control over the vehicle. It is necessary to guarantee system safety of the AD/ADAS application, which includes both "classic" functional safety according to ISO 26262 and specific areas like Safety of the Intended Functionality (SOTIF) and others. However, safety remains safety, that is, absence of unreasonable risk. All safety activities within a project, therefore, need to have their source in a Hazard Analysis and Risk Assessment (HARA), encompassing all relevant aspects, including operational situations, description of functionality and other parameters.

Already from the description a HARA for an AD/ADAS is going to be a complex task. Here we demonstrate an approach for complexity management of HARA for an ADAS system. A manageable overview of potential hazards resulting from malfunctions as well as from external causes was obtained and SOTIF validation goals were defined.

CITATION: Lurie, O. and Miller, J., "Hazard Analysis and Risk Assessment beyond ISO 26262: Management of Complexity via Restructuring of Risk-Generating Process," SAE Technical Paper 2018-01-1067, 2018, doi:10.4271/2018-01-1067.

Introduction

According to ISO 26262-3, a HARA consists of situation analysis and hazard identification, classification of hazardous events, and the determination of the safety level. The safety level largely determines further stages of the safety lifecycle, up to the safety validation.

Safety of the Intended Functionality (SOTIF) is an extension of the safety lifecycle. While the functional safety covers the hazards caused by malfunctioning behavior (i.e. unintended behavior of the item with respect to its design intent), DPAS 21448 "Safety of the Intended Functionality" [1] describes a specific safety lifecycle addressing performance limitations of the intended behavior or by reasonably foreseeable misuse by the user. This lifecycle includes "SOTIF HARA". We consider the SOTIF HARA to be an extension to the HARA according to ISO 26262 [2].

The following features distinguish the HARA related to the SOTIF from the HARA described in ISO 26262-3:

- *Hazard analysis:* even though the severity and controllability estimations use the same scales, their determination are specific for SOTIF hazards.

- *Safety levels* are not specifically addressed. The term "acceptable risk", which is often used in the document, refers to the acceptability of severity and controllability (S0 and C0 evaluations respectively).

- SOTIF HARA includes the *specification of a validation target*. Evidently, specification of a validation target requires the method of validation also to be specified.

The latter point may be solved in different ways. DPAS 21448, Annex B, suggests real-world driving tests on public roads. Once the cars equipped with the system in validation have driven the defined number of kilometers, their statistics can be compared to the available statistics of human driving, considering given region and type of a hazardous event. The GAMAB principle may be applied here: if the automated function performs at least as safe as human drivers, it may be considered safe.

According to ISO 26262, hazard identification is based on the situation analysis, which in turn consists of operational situations and the operating modes for the system in question. That means, a HARA consists of multiple situations often different only in minor details. This approach generates situations which may be different from each other only in minor details. Fitting the driving statistics into such detailed state description is not always possible or may be too arbitrary. Besides that, the management of many situations is problematic due to its complexity.

The reference HARA used for this paper contains 20 operating states to be considered for an AD system, which are analyzed in 28 situations. Out of 560 possible combinations, 166 were selected for further considerations. Those were counter-imposed to 23 possible environmental conditions. 1080 resulting situations are selected for final analysis. After the situations are clustered and possible malfunctions are added, the final analysis spreadsheet contains 1867 lines, each containing a pair of situation and a hazardous event.

In the following chapter, a method for lowering the HARA complexity by re-design of the underlining risk-generation process is described. An example of application for this approach is demonstrated. Its viability, strong and weak points are discussed. Finally, conclusion and an outlook over the future work is presented.

SOTIF HARA and State Space Explosion

State space explosion problem is a term coming from computer science, specifically from the area covering formal model checking of algorithms. Essentially, state space explosion is a situation where the number of states a system may assume (and therefore the states which should be considered by the model-checker) grows exponentially with increasing numbers of the parameters to be considered. By way of analogy, we may speak of the state space of the HARA, meaning all potentially hazardous situations which need to be analyzed.

HARA Composition

The goal of the HARA is to analyze hazards present in the system with regard to the system's different use cases. Therefore, the potentially hazardous situations of the HARA are built out of the superposition of those two. Below, we analyze the process for generation of hazards and of the use cases, i.e. the "risk-generating" process.

HAZARDS

According to ISO 26262, the HARA is performed on an *item*, i.e. a function implemented on the vehicle level is analyzed. Hazards are expressed as functional failure modes of the item. The failure modes of a technical system may be evaluated by analysis of its composition and working principles.

In case of SOTIF, the function in question is normally a part of driver's responsibilities which is being taken over by an AD/ADAS. The definition of this function should include driver's task breakdown (as the human drivers do not consciously distinguish between e.g. control tasks in longitudinal and lateral dynamics) and the sub-tasks which used to be solved by the driver and now going to be automated. The failure modes to be considered here are the "failure modes" of the driver, i.e. possible failures in fulfilling a driving-related task, as driver is now out of the loop. [3]. HAZOP uses matrix consisting of guidewords and signal names to generate a set of possible hazards [4]. The generated set is then subjected to plausibility analysis. Implausible combinations of signals and guide-words are then excluded. However, according to our experience, it is rarely possible to exclude driver-related hazards identified by HAZOP upfront, as drivers are prone to very different mistakes in controlling their vehicles. Errare humanum est.

USE CASES

Use cases for HARA include operating states, driving situations and environmental conditions. Not all environmental conditions are relevant for all states, e.g. for a passive safety system (i.e. a system minimizing harm to vehicle occupants after crash) it does not matter, whether the road surface was slippery before crash or not. However, in case of AD/ADAS functions, the object of the analysis is the driver's behavior. It is hard to predict which parameters are influencing the driving and which are not, therefore most of the parameters shall be taken into account. Besides that, an AD/ADAS function should be analyzed in all driving situations, both in those where it is intended to be used and in those, where it is not.

CHAPTER 5

HARA and the Hidden Semi-Markov Chain

Driving (as virtually any process) may be represented in the form of a transition diagram (see Figure 1). Furthermore, a Markov process can be built upon this diagram [5]. The process would describe the development of a hazardous event (e.g. an accident), i.e. precisely the events studied by the HARA.

The driver-dependent part of transitions shall be considered under "hazards" (more precisely, the "hazards" consider the situations when the driver or an ADAS/AD system is taking a wrong decision), while the driver-independent transitions shall be considered under "use cases". The probability of the use case constitutes the "exposure" (E) parameter of the ISO 26262-compliant HARA, while the probability of transition between hazards and "no accident" state defines "controllability" (C). Severity (S) is a property of the accident; it is not directly derivable from the semi-Markov chain pictured on the Figure 1.

The word "hidden" in the title of this section refers to the fact that there is no possibility to measure many of the probabilities and the rates of the chain directly. The major source of information on traffic statistics are the accident statistics, i.e. we are dealing here with a sample for which statistics are skewed by definition, as we would never know out of driving statistics how many people have experienced the hazard and managed to avoid the accident.

It is possible to calculate the distribution of the probability of the system being in each state of the semi-Markov chain, if the form of probability distributions for the transitions and its parameters are known. For the transition diagram presented on the Figure 1, the number of the transitions is the power of the following set:

$$Transitions = UC \times Hazards \cup Hazards \times Accidents \tag{1}$$

Each transition is governed by a specific probability distribution, which may differ from other distributions not only in parameters, but also in its form.

The approach of this paper is to find another, smaller set of probability distributions by changing the transition diagram of the semi-Markov processes underlying the HARA, while still being able to extract the HARA-relevant information as well as to validate the results against traffic and accident statistics and to generate a suggestion with regard to the validation of the SOTIF.

FIGURE 1   Semi-Markov chain representing processes studied via HARA.

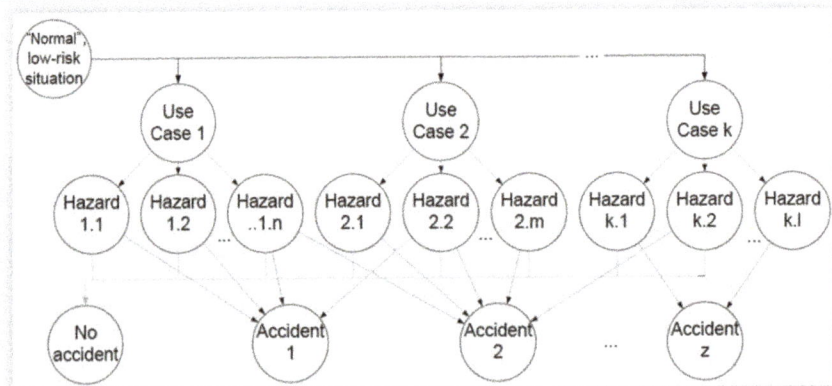

Restructuring of Risk-Generating Process

The road traffic is an extremely regulated area of human activity. It cannot be considered a Brownian motion. Road behavior of trained drivers features a combination of taught maneuvers under influence of the actual driving situation, which in turn is determined by the weather conditions, road type, behavior of other traffic participants and other parameters.

In order to minimize the number of the use cases, it makes sense to bound them to something easily identifiable and measurable. For AD/ADAS systems, it is beneficial to follow the approach proven by the process of teaching for human drivers, i.e. to benchmark the systems' performance in separate traffic situations. The choice of traffic situations as a milestone for risk development is substantiated by the relative accessibility of statistics on the use of different type of roads (cf [6]). Furthermore, accident statistics often gives information on the driving situation preceding the accident, allowing easier consideration of driving situations w.r.t. the causes of the accidents. The usage of driving statistics obtained from human drivers is valid as long as the cars equipped with AD systems represent a minority of vehicles on the roads. As long as there is no surge in the amount of AD-equipped vehicles, the actual driving statistics reflects the benchmark for the driving safety.

A model implementing driving situations as important predictors of the risk is shown on Figure 2. According to the model, it is assumed that initially (e.g. when car is stopped and parked), there is a situation which can be considered low risk or background risk. From here, the risk increases depending on the driving situation. In stating so, we *qualitatively* compare risks related to e.g. driving in a multi-lane road with no traffic with those related to driving on a single-lane road with vehicle parked alongside and tense traffic, and conclude that the latter has "higher risk" associated with it than the former. Our goal, however, is to estimate the risk *quantitatively via* probability of an accident, and to determine the tolerable risk, which is then used as a target for AD/ADAS validation. Therefore, we need a semi-Markov process as discussed above, and we obtain it by way of modification of the chain depicted on the Figure 1, taking into account the risk model.

Figure 3 depicts the transition diagram for the semi-Markov chain which could be used for HARA. For the transition diagram presented on the Figure 3, the number of the transitions is the power of the following set:

$$Transitions = Situations \cup Situations \times Accidents \qquad (2)$$

FIGURE 2 Model of road traffic risks.

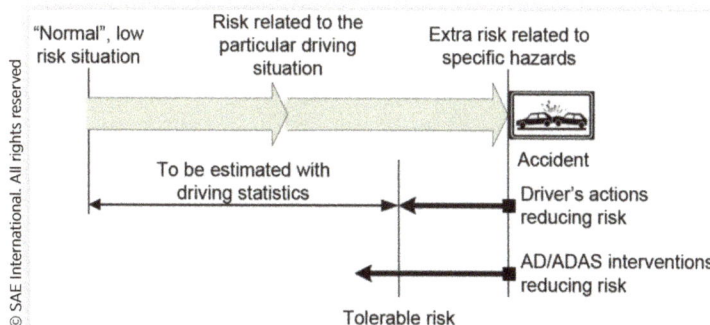

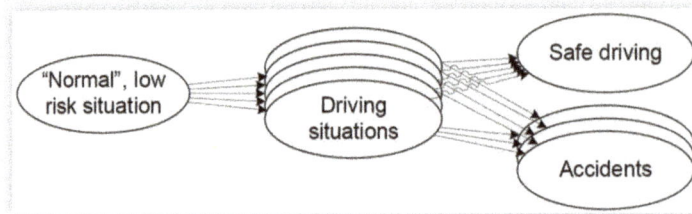

Comparing equations (1) and (2), it is safe to assume that the Figure 2 yields a semi-Markov process with considerably less transitions than the Figure 1.

Automatic Emergency Braking (AEB) Example

Markov Chain Solution

In order to illustrate the proposed method, a HARA for an AEB functionality will be studied.

Automatic Emergency Braking (AEB) detects an impending forward crash in time to avoid or mitigate the crash. These systems first alert the driver to take corrective action to avoid the crash. If the driver's response is not sufficient to avoid the crash, the AEB system may automatically apply the brakes to assist in preventing or reducing the severity of a crash [8]. AEB are mandated for some categories of trucks (see UN ECE R 131 or EU 661/2009). It is now being introduced also in passenger cars.

The HARA for AEB functionality is being performed under consideration of the two goals: first, ASIL shall be determined according to ISO 26262, and second, the driving distances for SOTIF validation shall be calculated. The HARA has been performed on the basis of the urban driving scenarios HARA (see [3]). The parameter of controllability by the driver was additionally validated by Monte-Carlo simulation [9]. SOTIF validation mileage was calculated via probabilistic approach.

The probabilistic approach included a stochastic process with multiple states modelled via state transition system as shown on Figure 3. For simplicity, all transitions rates were considered stationary; here we may speak of a Markov process instead of a semi-Markov process as considered above.

The sojourn probabilities in a Markov process are described by Chapman-Kolmogorov equation:

$$\dot{P} = \Lambda P \tag{3}$$

Where P is a vector representing sojourn probabilities and Λ is a transition matrix with an element λ_{ij} denoting transition rate between states i and j.

In the equation (3) we consider the probability against the number of driven kilometers $P(l)$, with $\dot{P} = dP/dl$.

Eq. (3) possesses an analytical solution [5]:

$$P_l = \exp(\Lambda l) P_0 \tag{4}$$

Where P_l is the sojourn probability vector after l kilometers of travel, and P_0 is the initial condition for the same. $P_0 = [1\ 0\ 0\ \ldots\ 0]^T$ by definition, as all vehicles always start from the "normal, low risk" state. Driving statistics defines l (average yearly mileage of one vehicle) and Pl (vector containing probabilities of different accidents happening in a year). Solving the eq. (4) for Λ will yield all transition probabilities between states depicted in Figure 3; distance to accident may be calculated as an inverse of the relevant transition rate:

$$MDTA_{ij} = 1 / \lambda_{Situation_I \to Accidents_J} \tag{5}$$

Eq. (5) shows that if Λ is known, it is easy to find the expected time to a specific accident in each driving situation. The validation can be performed focused on the driving situation via driving for $MDTA_{IJ}$ kilometers in a driving situation I and proving that no breach of safety has happened. A safety coefficient may be used as shown in PAS 21448 Annex D.

Eq. (4) has an analytical solution for Λ involving matrix logarithm. Matrix logarithm involves calculation of a sum of a row which converges only conditionally (in contrast, the row representing the matrix exponent always converges). Therefore, numerical solution was chosen involving random gradient descent to find Λ.

Regions of the Transition Matrix

The transition matrix Λ may be divided into the regions as shown by the eq. (6):

$$\Lambda = \begin{bmatrix} O_1 & D & O_2 \\ O_3 & O_4 & A \end{bmatrix} \tag{6}$$

Here D represents the probabilities of transition from the initial "low risk" state into the driving state, A represents the transitions between driving states and the accidents (including "safe driving" as non-accident), and O are zero matrices of the appropriate size. The D part is taken from the traffic data [7]. Our goal is to determine A fulfilling the Eq. (4).

The matrix A itself is also sparse, as transitions between driving situations and accidents are not arbitrary. Table 1 shows existing transitions (non-zero transition rates) in the matrix A.

Kinds of the accident are selected according to [7]:

1. Collision with another vehicle which starts, stops or is stationary
2. Collision with another vehicle moving ahead or waiting
3. Collision with another vehicle moving laterally in the same direction
4. Collision with another oncoming vehicle
5. Collision with another vehicle which turns into or crosses a road
6. Collision between vehicle and pedestrian
7. Collision with an obstacle in the carriageway
8. Leaving the carriageway to the right or left

Driving situation according to HARA (cf [3]):

S1. Vehicle in motion at constant speed, in heavy traffic
S2. Vehicle accelerating, in heavy traffic
S3. Vehicle changing lanes on a multi-lane road with heavy traffic
S4. Vehicle turning, on an extreme curve or at a marked or unmarked intersection, with heavy driving

TABLE 1 Non-zero transitions in the part A of transition matrix Λ

Kind of accident	S1	S2	S3	S4
1		P	P	
2		X	X	
3			X	P
4				P
5		P	P	P	P
6		X	X		X
7		X	X		X
8					P

TABLE 2 Median Distance to Accident (MDTA), thousands of km

Situation	S	MDTA, ×1000 km
Vehicle in motion at constant speed, in heavy traffic	S1	60
Vehicle accelerating, in heavy traffic	S2	75
Vehicle changing lanes on a multi-lane road with heavy traffic	S3	60
Vehicle turning, on an extreme curve or at a marked or unmarked intersection, with heavy driving	S4	6
Total	$\sum S_{1..4}$	201

In the Table 1, both "X" and "P" denote a non-zero element of the matrix A. The difference here is assumed in the magnitude of the transition rates. "P" (meaning "potentially") marks transitions which are possible only under the condition of a breach of the traffic code. Consider "P" in row 1, column S1. This combination refers to a vehicle, being in motion with constant speed in heavy traffic, colliding with a vehicle driving off its parking spot. According to the traffic code, the driving off vehicle shall let the traffic pass before leaving the spot. For this example, we assume that transitions related to the breach of the traffic code happen at least one order of magnitude more rarely than the "normal" transitions (this assumption may be revised or completely ignored later, as it is not important for the method).

Table 2 shows the median distances to accident depending on the driving situation. The data are obtained based on the transition rates of a Markov chain depicted on Figure 3, using Eq. (5).

The total mileage presented in Table 2 is comparable with the mileage calculated via alternative methods (cf. [9] where 153,000 km is the calculated MDTA for all driving states based on the US driving statistics).

Is There Another Way to Do It?

Alternatively, the validation distance may be calculated applying the formula from PAS 21448, Annex B (Eq. 7) directly.

$$D_{km} = \frac{NK}{A_v} C_a P \tag{7}$$

Here N is the number of vehicle in the field; K - average annual mileage per vehicle (i.e. NK is the average mileage all vehicles travel per year); A_v is the amount of accidents of the relevant type per year (for AEB it would be rear- and front-end collisions); C_a is the assurance factor; and P is the probability of accident resulting from the given situations.

Calculation of P poses a specific challenge to this approach. It was suggested to use Monte Carlo modelling to determine P [10]. However, the modelling requires validation, which due to granularity of use cases cannot be performed on the available driving statistics. Having to validate 1867 sets of model parameters further complicates the problem of calculation of validation mileage in SOTIF HARA.

Summary

Within this work, we have set out the goal to reduce the complexity of SOTIF HARA while still being able to define validation goals. This we started with a HARA containing 1867 lines (about 180 of those relevant for AEB functionality). We ended up with a model for calculation of the SOTIF validation range, with 32 relevant states (4 driving situations vs. 8 possible collision models), revealing almost 6-fold reduction. All the rest of the required information was obtained via driving statistics and other reference documents.

Besides the calculation of SOTIF validation targets, the Markov-based stochastic approach described above may be used for estimation of any other safety-related parameter, incl. exposure and controllability parameters.

Outlook

A simple example presented here considers a fully Markov chain with stationary transition rates, leading to considerable limitations. Based on the transition model introduced on Figure 3, semi-Markov chain may be built representing various distributions of transitions and states. Those distribution may be further parametrized, allowing the better fitting and validation of the model.

Another prospect for improvement of the approach is using better, more precise data. Here, a yearbook on traffic accident statistics from German Federal Office for Statistics was used as data source. Further sources like GIDAS study may provide for better model validation and more precise parametrization.

Contact Information

The main author:
Oleg Lurie
Safety Analyst at ZF TRW
Global Engineering Excellence - Systems Safety
TRW Automotive GmbH
Fritz-Reichle-Ring 8
D-78315 Radolfzell/Germany
Telefon/Phone +49.77732.939.1467
Telefax/Fax +49.7732.939.1820
oleg.lurie@zf.com

Definitions/Abbreviations

HARA - Hazard Analysis and Risk Assessment
AD - Autonomous Driving
ADAS - Advanced Driver Assistant System
SOTIF - Safety of the Intended Functionality
GAMAB - Performing globally at least as good as
Eq. - Equation

References

1. ISO/WD PAS 21448, "Road Vehicles - Safety of the Intended Functionality," ISO Working Draft, 2013.

2. Lurie, O., "Where Lifecycle Starts and Ends," presented at *the Conference Operational Safe Vehicles for Automated Driving*, Berlin, Germany, September 19-20, 2017.

3. Koopman, P. and Wagner, M., "Challenges in Autonomous Vehicle Testing and Validation," *SAE Int. J. Trans. Safety* 4, no. 1 (2016):15-24, doi:10.4271/2016-01-0128.

4. IEC 61882:2016, "Hazard and Operability Studies (HAZOP Studies): Application Guide," IEC Standard, Rev., March 2016.

5. Birolini, A., *Reliability Engineering: Theory and Practice* (Berlin/Heidelberg/New York: Springer, 2009), ISBN:3-540-63310-3.

6. VDA 702, "Situations Catalogue for E Parameters according to ISO 26262-3" (in German), VDA Standard, Rev., June 2015.

7. "Traffic. Traffic Accidents" (in German), Federal Statistical Office of Germany, Wiesbaden, 2016.

8. https://www.safercar.gov/Vehicle-Shoppers/Safety-Technology/AEB/aeb.

9. Fabris, S., "ADAS Design according to ISO 26262 and SOTIF Principles," presented at *IMECH 2017 Conference*, Birmingham, May 2017.

10. Priddy, J., Harris, A., and Fabris, S., "Method for Hazard Severity Assessment for the Case of Undemanded Deceleration," presented at *VDA Automotive System Conference*, Berlin, 2012.

CHAPTER 6

The Science of Testing: An Automotive Perspective

Siddartha Khastgir, Stewart Birrell, Gunwant Dhadyalla, and Paul Jennings
University of Warwick

Increasing automation in the automotive systems has re-focused the industry's attention on verification and validation methods and especially on the development of test scenarios. The complex nature of Advanced Driver Assistance Systems (ADASs) and Automated Driving (AD) systems warrant the adoption of new and innovative means of evaluating and establishing the safety of such systems. In this paper, the authors discuss the results from a semi-structured interview study, which involved interviewing ADAS and AD experts across the industry supply chain.

Eighteen experts (each with over 10 years' of experience in testing and development of automotive systems) from different countries were interviewed on two themes: test methods and test scenarios. Each of the themes had three guiding questions which had some follow-up questions. The interviews were transcribed and a thematic analysis via coding was conducted on the transcripts. A two-stage coding analysis process was done to first identify codes from the transcripts and subsequently, the codes were grouped into categories.

The analysis of transcripts for the question about the biggest challenge in the area of test methods revealed two specific themes. Firstly, the definition of pass/fail criteria and secondly the quality of requirements

CITATION: Khastgir, S., Birrell, S., Dhadyalla, G., and Jennings, P., "The Science of Testing: An Automotive Perspective," SAE Technical Paper 2018-01-1070, 2018, doi:10.4271/2018-01-1070.

(completeness and consistency). The analysis of the questions on test scenarios revealed that "good" scenario is one that is able to test a safety goal and ways in which a system may fail. Based on the analysis of the transcripts, the authors propose two types of testing for ADAS and AD systems: Requirements-Based Testing (traditional method) and Hazard Based Testing. The proposed approach not only generates test scenarios for testing how the system works, but also how the system may fail.

Introduction

The introduction of Advanced Driving Assistance Systems (ADASs) and Automated Driving (AD) systems in cars have many benefits ranging from increased safety [1, 2], lower emissions, reduced traffic congestion [3, 4] and more useful time for the driver [5]. The potential benefits of automated systems have led the push towards their commercialisation. Interestingly, the public opinion about *"completely self-driving (fully automated) vehicles"* has been shown to be in line with the proposed safety benefits [6]. Automated systems offer many benefits in other industries too where they have been introduced, e.g. aviation, nuclear, chemical process, railways etc. Unfortunately, the introduction of automation in these industries was coupled with many accidents, some of which have continued to repeat themselves [7]. Even within the automotive industry, many relatively advanced features (at the time), have caused vehicle re-calls due to faulty software, costing millions of dollars to the manufacturers; e.g. to fix the ignition switch issue, General Motors spent approximately $400 million for the 2.6 million affected vehicles [8]. Fixing a bug during the development process costs an average of $25, while after release it increases to $16000 on an average [9]. A bug in a released product could be caused due to: 1) incorrect requirements 2) missing requirements 3) release of untested code, 4) testing sequence differs from use sequence 5) user applied untested input values 6) untested operating conditions [10]. The latter was illustrated in the Ariane 5 disaster [11], where software was reused from Ariane 4 software in the Ariane 5 system without enough testing [12]. This importance of operating environment and potential consequence of untested inputs was also seen in the recent Tesla "Auto-pilot" system crash [13]. It has been suggested that majority of the software related accidents are a result of the operation of the software rather than its lack of operation [14].

Therefore, in order to realize the benefits of automation or any other system, we need to ensure that the systems have a safe and a robust functionality. This may be achieved by testing and certification of the systems. However, lack of standardized test methods and test scenarios; and the lack of international standards to define safety requirements for automated systems, have led to a subjective interpretation of "safety", particularly for ADAS and AD systems in vehicles. While the ISO 26262 standard [15], provides some guidance for testing methods and approaches for a product development cycle, it too falls short to deal with the complexities of ADAS and AD systems. Furthermore, even with ISO 26262 - 2011 been increasingly adopted in the industry, there is still a lack of a *"quantified and rigorous process for automotive certification"* [16]. This is caused due to the lack of objective quantification of severity, exposure and controllability ratings which comprise the ASIL rating, causing inter- and intra-rater variations [16, 17].

Current luxury cars are a complex system with over 100 millions lines of code as compared to 7 million in a Boeing 747 airplane [18]. The introduction of ADASs and

AD systems is going to further increase the complexity many fold with multiple interactions between subsystems. Additionally, ADASs and AD systems offer a new challenge for testing and safety analysis [19]. While a variety of ADASs and AD systems exist or are in development, each of them offers a different kind of a challenge for testing. The move towards higher levels of automation is coupled with the challenge of testing and safety analysis as it needs complex solutions to include interactions between a larger number of variables and the environment. It is suggested that in order to prove that automated vehicles are safer than human drivers, they will need to be driven for more than 11 billion miles [20]. Even after 11 billion miles, such testing will *"only assure safety but not always ensure it"* [21], thus suggesting vehicle level testing or real world testing before start of production (SOP) would not be enough to prove safety of the automated driving systems [22, 23]. While software testing has been said to be the *"least understood part of the (system) development process"* [10], the authors believe that a scientific approach needs to be adopted to solve the challenge of identifying scenarios that capture the complex interactions within systems and system-environment in an efficient manner.

Understanding Scenarios

These complex interactions can be captured as use cases which *"describe the system behaviour as a sequence of actions linking the result to a particular actor"* (e.g. driver). Subsequently, scenarios (a specific sequence of a use case) present possible ways in which a system may be used to accomplish a desired function. However, writing scenarios require detailed domain knowledge, which is only found with experts. Moreover, the term "use cases" and "scenarios" have been used with a fuzzy meaning [24, 25]. A use case is a collection of scenarios bound together by a common goal [25] and implies *"the way in which a user uses a system"* [26]. Scenarios have been suggested to have at least four different meanings: 1) scenarios to illustrate the system 2) scenarios for evaluation 3) scenarios for design 4) scenarios to test theories [24]. It is worth elucidating that a scenario that is good for illustrating a system demo (i.e. demonstration) may not be good for evaluating the basic functions (i.e. requirement based testing), as the former only uses a limited number of examples. Similarly, scenarios to test theories establish the strengths and more importantly the weaknesses of a design. Therefore, they go beyond the traditional requirement based testing.

Existing Requirements Based Testing (RBT) approach widely used in the industry, only ensures that the system meets its requirements while failing to identify the exceptions explicitly. Some exceptions may be covered sporadically due to the experience of historic failures rather than a scientific approach. Additionally, RBT is not able to ensure completeness of requirements. Requirements reflect the expert's view of system's functionality and possible usage. The identification of the requirements has a degree of subjectivity associated with it [27]. Different experts with different background knowledge analyse and classify systems differently, leading to an inter-rater variation in understanding requirements [17, 28].

This is evident in the variation in the classification and identification of scenarios like the *"Black Swan"* scenarios or the *"unknown unknowns"* (scenarios that we do not know that we do not know) associated with the functionality of the system [29]. While requirements based testing captures the *"known knowns"* efficiently, the inability to ensure its completeness leads to the occurrence of *"unknown knowns"*, *"known unknowns"* and the Black Swan scenarios. In addition, to avoid the variation in understanding of the terms use-cases, scenarios and test cases, the authors adopt the definition as described in [30].

FIGURE 1 Methods for testing functional safety and technical requirements as per ISO 26262 - 2011 Part 4.

Methods	ASIL			
	A	B	C	D
1a	Requirement-based test[a]	++	++	++	++
1b	Fault injection test[b]	+	+	++	++
1c	Back-to-back test[c]	o	+	+	++

FIGURE 2 Methods for deriving test cases for software unit testing as per ISO 26262 - 2011 Part 6.

Methods	ASIL			
	A	B	C	D
1a	Analysis of requirements	++	++	++	++
1b	Generation and analysis of equivalence classes[a]	+	++	++	++
1c	Analysis of boundary values[b]	+	++	++	++
1d	Error guessing[c]	+	+	+	+

Types of Testing

The international standard ISO 26262 [15] is the automotive industry best practice standard for functional safety. ISO 26262 Part 4 and Part 6 provide guidance on different methods for testing and for deriving test cases for software integration testing and software unit testing respectively. ISO 26262 - 2011 Part 4 and Part 6 recommend the use of test methods like requirement based test, fault-injection test and back-to-back comparison test (Figure 1). For each of the test methods, the standard recommends methods like analysis of requirements, analysis of equivalence classes, analysis of boundary values and error guessing as methods for deriving test cases (Figure 2).

The ISO 26262 standard also recommends some metrics to measure the completeness of the testing process. These include coverage metrics like function coverage and call coverage at the integration level (Figure 3) and branch coverage, statement coverage and MC/DC (Modified Condition/Decision Coverage) at unit level (Figure 4).

As suggested by ISO 26262 - 2011, testers tend to go all-out for coverage metrics. While this is "a metric", it needs to be highlighted that it should be treated as the minimum metric. If achieving high coverage was the golden bullet for testing, then products in use would have very few bugs [10]. Since there are infinite possibilities for input suite, testers tend to use the *"best"* sample test to *"adequately"* test the system, where "best" and "adequately" is based on the subjective judgement of the tester [10].

FIGURE 3 Software architecture level structural coverage metrics per ISO 26262 - 2011 Part 6.

Methods	ASIL			
	A	B	C	D
1a	Function coverage[a]	+	+	++	++
1b	Call coverage[b]	+	+	++	++

FIGURE 4 Software unit structural coverage metrics as per ISO 26262 - 2011 Part 6.

Methods	ASIL			
	A	B	C	D
1a	Statement coverage	++	++	+	+
1b	Branch coverage	+	++	++	++
1c	MC/DC (Modified Condition/Decision Coverage)	+	+	+	++

While the advent of automated systems in automobiles has led to an increasing focus on incorporating functional safety in the design process, the current version of the standard fails to provide guidance on systems with high automation.

The industry, acknowledging this gap has attempted to address it with the upcoming SOTIF (Safety Of The Intended Functionality) publically accepted specification [31]. This paper captures the essence of the gap in knowledge of testing for ADAS and AD systems and proposes a means to fill this gap.

Methodology

In order to understand the testing approach being undertaken by the automotive industry towards ADAS and AD systems to uncover the *"unknown unknown"*, *"unknown known"* and *"known unknown"* scenarios, the authors conducted a semi-structured interview study involving verification and validation experts in the automotive domain. Semi-structured interviews were conducted to understand the existing knowledge base for test scenario generation process in the automotive industry and their understanding and expectations from a good/ideal test scenario. Semi-structured interviews were adopted as they provide the flexibility to the interviewee to provide wider information and thus richer data, by enabling the formation of a understanding between the interviewer and the interviewee due to face to face contact [32]. Additionally, they allow the flexibility to examine topics in different degrees of depth (as per interviewees' interest and background) [33]. The interviews were transcribed and the text was sanitized to remove any proprietary mentions. A coding analysis was performed on the sanitized text and themes and categories were identified from the various interview answers. Coding analysis groups participant responses which are similar to give a broader understanding of responses.

In order to prevent any bias, the interviewees were allowed to talk freely while answering the questions and were not prompted for any answers. Participant interviews were transcribed into text and were later coded to perform thematic analysis. Key themes were identified in both parts of the interview.

Ethical approval for the study was secured from the University of Warwick's Biomedical & Scientific Research Ethics Committee (BSREC). All interview transcripts were anonymized and stored in a secure location and University of Warwick's data handling procedures were followed.

Participants

Eighteen industry experts, each participant having over 10 years' of experience in the field of testing and development of systems in the automotive industry were recruited for this study. Participants were selected from a diverse demography cutting across the automotive supply chain. Nine participants represented OEMs (Original Equipment Manufacturers), eight participants represented Tier 1/2 suppliers and the remaining participant represented academic /research organizations / start-ups working in the area of automated driving. To ensure independence of the interviewees, participants were recruited from different countries including the UK, Germany, India, Sweden, Japan and USA. The interviews lasted between 28.63 minutes and 103.15 minutes (average interview length: 48.25 minutes). Interviewees were also assured that any of the responses will not be identifiable to them as the transcripts would be anonymized before they were analysed.

TABLE 1 Interview question design (follow-up questions)

Guiding Ques. #	Follow-up Question(s)
1	Reason for selecting a test method? What tools do you use as a part of your test setup?
2	What is your biggest challenge?
3	How was the metric developed? Is it a standard metric? (company internal or industry standard)
4	What test scenarios do you use while doing real world / virtual testing?
5	What aspects are critical while developing a test scenario for autonomous system?
6	How did you develop those (for good quality test scenario) criteria?

Interview Questions Design

The interview was structured with six guiding questions, which were divided into two themes: 1) test methods (three questions) 2) test scenarios (three questions). Each guiding question had a set of follow-up questions, which were asked depending on the content of the answers. The set of follow-up (prompting) questions are described in Table 1. The follow-up questions were used to aid participants thought process and were designed to be minimally prescriptive to avoid biasing the answers. The guiding questions were formulated by the existing gaps in the literature. The six guiding questions were the following:

Test Methods

1. What test methods do you use for testing of automotive systems?
2. What are the challenges for each test method that you have faced?
3. What metric do you use to measure sign-off criteria for testing automated systems?

Test Scenarios

4. How do you ensure robust testing of automated automotive systems in various driving conditions?
5. How do you develop test scenarios for testing automated systems?
6. What criteria do you think make a good quality test scenario?

Data Analysis

As this study employed a semi-structured interview format, the analysis of the data was mostly qualitative. In order to structure the data analysis and identify trends in the collected data, a coding strategy was used. A code *"is a word or short phrase that symbolically assigns a summative, salient, essence-capturing , and/or evocative attribute to a portion of language-based or visual data"* [34]. By reading through the transcribed interview text, codes were assigned to the text which enabled conversion of the interview text into an easy to understand tabulated format. An example of a code and corresponding text is discussed here. One of the responses to the question on the biggest challenge in testing faced by the interviewees was, *"it is difficult to create the specification to verify against and because of the lack of specification, it is difficult to put a criteria for completeness of testing".* The corresponding code assigned to the text was *"how to ensure completeness of requirements"* and *"how to judge test completeness".* However, it is evident that such a coding process is subjective due to the understanding and biases of the coder. In order to overcome this, a two stage coding process was followed which was reviewed by an independent expert.

First Cycle Coding

The first coding cycle involved reading through the interview transcripts to assign codes. As the data in a semi-structured interview transcripts can be varied, different methods of coding such as structural coding, descriptive coding, process and in-vivo coding, were used [34].

Second Cycle Coding and Category Identification

Since different coding methods were used in the first coding cycle, some of the codes were similar or split. In order to synthesize the first cycle codes to develop a more cohesive understanding, axial coding was used in the second phase which led to the creation of categories for the first cycle codes. Table 2 illustrates the coding process for the answers received to question 5.

TABLE 2 Development of codes for question 5

Participant answers	1st cycle codes	2nd cycle codes
"what kind of environmental influences could lead to an ill function"	Environmental factors, failures	
"try to define a test to see the degradation of the performance"	Degraded performance, faults	Identify failures, system limits, and hazards.
"understand situation in which our system will reach any kind of limit"	System limits, degraded performance	
"fidelity of test scenario comes down to the FMEA. Because out of the FMEA there is possibility of a failure you need a control method for"	Identify failures, systematic way, FMEA	using **systematic method** to identify failures, hazards
"we would engineer faults into the system.... Blocking the radar. Put radar absorbent material (RAM) for the radar."	create faults, block sensors	
"it is currently done via FMEA, System FMEA and Hazard Analysis"	Systematic way, FMEA, HARA	using **systematic method** to create test scenario library
"have a catalogue of tests"	Test library	
"systematic way (of) what kind of influencer I have into the behaviour of the functionality..."	Factors influencing functionality, systematic method	
"you can think about you have a matrix... then we look at what kind of combinations are possible"	Test library	

Results

While it was found that tools (software platforms) for test execution were not an issue for most organizations, the infrastructure requirement for test platforms (hardware-in-the-loop setup and instrumented test vehicles for real-world testing) had exponentially increased with ADAS and AD systems as compared to traditional automotive systems. In addition, the large amount of data handling required for sensors used in the ADAS and AD systems was another challenge.

In response to the first question on test methods used for testing, the participant responses could be grouped in two themes. One group of participants commented that they follow the software development V cycle and implemented model-based design tools using simulation in a major part of their development process. On the contrary the other group was of the opinion that simulation is of limited use for ADASs and AD systems as it is *"almost impossible"* to model sensors, especially RADAR and LiDAR sensors and they mostly depended on real world testing.

More importantly, the input to the test execution platform (test case vectors) was a common concern acknowledged by all participants. When asked about the biggest challenge faced by the participants while performing testing, two specific themes emerged. While the OEMs credited *"test case generation and definition of pass/fail criteria"* as their biggest challenge; tier 1/2 suppliers credited *"quality of requirements (including completeness and consistency)"* as their biggest challenge. This difference can be credited to the culture in the automotive supply chain where the suppliers develop individual systems and the responsibility for integration of these systems lies with the OEMs. However, both the groups failed to mention any solutions to the challenges faced by them during the testing phase; the ability to identify and define the *" known unknown"* and the *"unknown unknown"* scenario space.

When asked about the parameters and criteria for good test scenarios, there seem to be an agreement on the ability to test *"known unknown"* and *"unknown unknown"* situations, as a key feature of a good test scenario. However, a deeper analysis of the responses revealed two distinct themes on ways to achieve "good" test scenarios. Firstly, creating "good" scenarios from requirements is dependent on the skill and experience of the test specifiers. Secondly, "good" scenarios should be able to test safety goals and ways in which the system may fail or reach system limits. This is generally not covered by system requirements. Moreover, the need for a systematic method of identifying the system limits or failure scenarios was highlighted by the participants. Most experts mentioned that Requirements Based Testing (RBT) is insufficient as there is a challenge in ensuring completeness of requirements. RBT captures the typical scenarios as suggested by the requirements and represents the most common real world scenarios. Such testing ensures that the most common bugs are identified [10].

While approaches to improve requirement based testing have been discussed in literature [35, 36], discussion on the ability to increase the *"known known"* by identifying the unknown space is limited. One of the reasons mentioned by experts about RBT was that it is impractical to have a requirements document capturing the multitude of scenarios an automated driving system might encounter, rendering the classical V-cycle for software development obsolete.

In the testing process, it is important to establish when to stop testing and sign-off the system-under-test. When the participants were asked about a metric used to measure the sign-off criteria, to our surprise, the answers demonstrated the lack of any standard metric in place. Unfortunately, the sign-off point was dependent on the budget allocated and SOP time. However, all participants acknowledged that this wasn't the ideal situation and needs to change for ADASs and ADS systems. However, some participants did provide some insight into an ideal situation and using false positive and false negative rates as metric for sign-off.

When asked about how participants ensured that the ADASs and AD systems were tested robustly, they mentioned using a test catalogue which was developed from experience. However, all participants agreed that for ADAS and AD systems, more real world testing is needed due to challenges in simulation environment. On the time split between real-world and virtual testing, one of the participants commented: *"95% is real world testing and 5% is simulation. But for me it should 50-50. For the moment the robust model of the simulation is stopping (this to happen)".*

Discussion

One of the challenges of identifying *"black swan"* scenarios is their lack of correlation with time [23]. Based on the analysis of the interviews, to increase the area covered by the *"known knowns"* in the test scenario space, the authors propose a two-pronged approach to testing of ADASs and AD systems to create test scenarios and test cases (Figure 5). The first branch concerns using traditional RBT approach, while the second branch uses a Hazard Based Testing (HBT) approach for creating test scenarios. Traditional RBT method covers only a fraction of the possible test scenario space for the systems (Figure 5). The addition of the second testing branch (HBT) improves the coverage of the test scenario space by increasing the *"known known"* scenario space. However, it does not guarantee full coverage of the test space (Figure 5). While RBT checks the working of the system as per expectations (defined requirements), HBT explores how the system may fail by identifying possible failure scenarios.

HBT draws its inspiration from the world of security analysis. In security analysis, the use of misuse cases has been suggested as a way of testing for security concerns [37]. Misuse cases can *"help document negative scenarios"* [37]. The key to the success of HBT is to have a structured, robust and well-documented method of identifying hazards. This was also highlighted in the themes obtained from the analysis of the interview transcripts. The two themes were: *" failure or hazard scenarios"* and *"systematic method (objective) to obtain them"*. On being asked about how to develop test scenarios, one of the participants commented: *"try to define a test to see the degradation of the performance"*, while another participant mentioned: *"what kind of environmental influences could lead to an ill function"*.

In order to identify hazards, various methods like HAZOP, FMEA [38], Event Tree Analysis, JANUS [39], Accimaps [40], HFACS [41, 42, 43], Fault-tree analysis [44, 45], bow-tie analysis [46], System Theoretic Process Analysis (STPA) / Systems

FIGURE 5 Proposed testing approach for test-scenario generation.

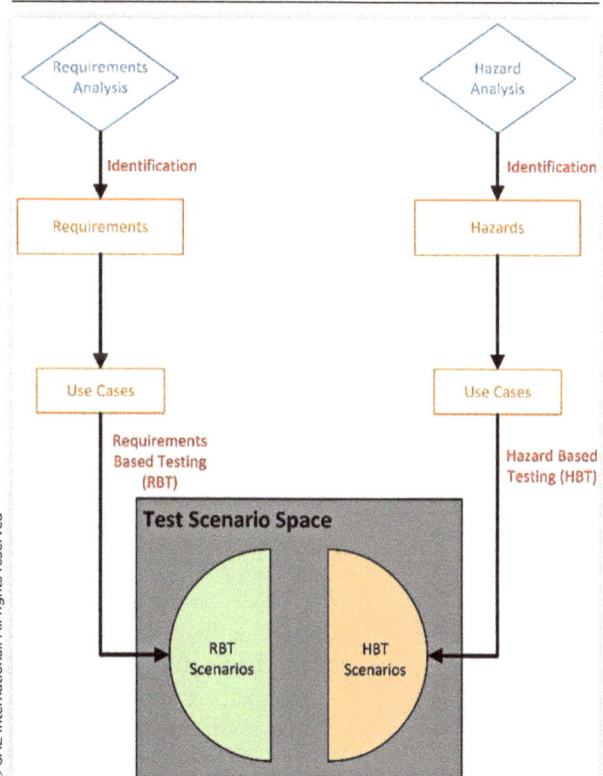

Theoretic Accident Model & Processes (STAMP) [47, 48, 49] etc. have been used in the industry and research community. Some of these methods were developed for simple systems and fall short in analysing modern ADAS and AD systems which have multiple interactions between system, human operator and the software [50]. In case of an Adaptive Cruise Control (ACC) system, rather than testing the functional requirements, more emphasis needs to be laid on identifying the hazards associated with the usage of an ACC system. An analysis of the ACC system using any of the earlier said methods to identify hazards would lead to one of the potential hazards as *"unintended braking"*.

Some of the hazard identification methods developed specifically for ADAS and AD system (e.g. HFACS, JANUS, STPA) further analyse the system interactions to identify that one of the potential causes of an *"unintended braking (hazard)"* could be the "vehicle maneuvering through a steep bend" causing the radar system to believe that there is an obstacle in front. Therefore an HBT approach would identify such situations which would have been missed in a traditional RBT approach.

In order to identify the safety goals and the hazards, a Hazard Analysis and Risk Assessment (HARA) process needs to be conducted. The automotive HARA has its own issues like subjective variation due to skill and experience of the testers and completeness of the HARA, some of these issues have been answered in the literature [17]. Once the systems have been tested, their capability and safe performance can be correctly established and can form part of the knowledge to be imparted to the drivers, in real time or before they start their usage, establishing their "informed safety" level to improve trust in ADASs and AD systems [51]. However, in order to create the "informed safety" level, a more systematic and structured process needs to be adopted to testing. As one of the interviewees mentioned, *"Testing is a science"*.

In this study, the authors had a limited sample size for the interview pool due to resource constraints. While a large number of practitioners are involved in the field of verification and validation, it would be a major challenge to interview a representative sample size. However, due to the expertise of the interviewees, the authors believe that the current findings provide an important insight in the future direction for testing of automated automotive systems, which will be an essential component of the system development process.

Conclusion

The lack of a scientific approach to testing has led to the inability of the industry to tackle the challenges offered by ADAS and AD systems in terms of testing. The authors interviewed 18 automotive experts, each with over 10 years' of experience in testing and development of automotive systems. "Creating test scenarios" and "ensuring completeness of requirements" were highlighted as the main challenges for testing of ADAS and AD systems. However, none of the experts could provide a solution to any of the two challenges.

Moreover, the experts suggested that a "good" test scenario is one that tests how the system fails or reaches its limits, in addition to having a structured approach to define the test scenario. Requirement based testing tends to elude capturing this test space, and thus according to the experts is not enough when it comes to testing ADASs and AD systems.

Based on a detailed analysis of the interview transcripts, the authors have proposed a new approach for testing to increase the coverage of the test scenario space. This has been achieved by reducing the occurrence of *"Black Swan"* scenarios (i.e., unknown unknowns) and *"known unknowns"*, by increasing the *"known knowns"* of the system. The proposed method comprises of a two-pronged approach to identifying test scenarios. The first branch comprises of the traditional requirement based testing (RBT) method, while the second branch comprises of a hazard based testing (HBT) approach. The latter requires the identification of hazards for a system. The *"known unknowns"*, *"unknown knowns"* and *"unknown unknowns"* get uncovered in the Hazard Based Testing branch. Safety goals, which give rise to hazards, are identified by conducting a Hazard Analysis and Risk Assessment (HARA) process. The proposed approach not only generates test scenarios for testing how the system works, but also how the system may fail, thus increasing the test scenario space.

Contact Information

Siddartha Khastgir
WMG, University of Warwick, UK
Address: International Digital Laboratory, WMG
University of Warwick, UK, CV4 7AL
S.Khastgir@warwick.ac.uk

Stewart Birrell
WMG, University of Warwick, UK
Address: International Digital Laboratory, WMG
University of Warwick, UK, CV4 7AL
S.Birrell@warwick.ac.uk

Gunwant Dhadyalla
WMG, University of Warwick, UK
Address: International Digital Laboratory, WMG
University of Warwick, UK, CV4 7AL
G.Dhadyalla@warwick.ac.uk

Paul Jennings
WMG, University of Warwick, UK
Address: International Digital Laboratory, WMG
University of Warwick, UK, CV4 7AL
Paul.Jennings@warwick.ac.uk

Acknowledgments

This work has been carried out under the VVIL project (Validation of Vehicle-in-the-Loop), funded by Innovate UK under the grant ref. 132302 and is a collaboration between Vertizan Limited and WMG, University of Warwick, UK. The authors would like to thank EPSRC (Grant EP/K011618/1) and WMG centre HVM Catapult, for providing the necessary infrastructure for carrying out this study. WMG is one of the seven centres that together comprise the High Value Manufacturing Catapult in the UK.

References

1. Carbaugh, J., Godbole, D.N., and Sengupta, R., "Safety and Capacity Analysis of Automated and Manual Highway Systems," *Transp. Res. Part C Emerg. Technol.* 6, no. 1-2 (1998): 69-99, doi:10.1016/S0968-090X(98)00009-6.

2.   Fagnant, D.J. and Kockelman, K., *Preparing a Nation for Autonomous Vehicles: Opportunities, Barriers and Policy Recommendations*, 3rd ed. (SAGE, 2015), 0965-8564, doi:10.1016/j.tra.2015.04.003.

3.   van Arem, B., Cornelie, J.G.V.D., and Visser, R., "The Impact of Co-Operative Adaptive Cruise Control on Traffic Flow Characteristics," *IEEE Trans. Intell. Transp. Syst.* 7, no. 4 (2005): 429-436.

4.   Shladover, S.E., "Cooperative (Rather than Autonomous) Vehicle-Highway Automation Systems," *IEEE Intell. Transp. Syst. Mag.* 1, no. 1 (2009): 10-19, doi:10.1109/MITS.2009.932716.

5.   Le Vine, S., Zolfaghari, A., and Polak, J., "Autonomous Cars: The Tension between Occupant Experience and Intersection Capacity," *Transp. Res. Part C Emerg. Technol.* 52 (2015): 1-14, doi:10.1016/j.trc.2015.01.002.

6.   Schoettle, B. and Sivak, M., "Public Opinion About Self-Driving Vehicles in China, India, Japan, The U.S., The U.K. and Australia," ISBN UMTRI-2014-21, 2014, doi:UMTRI-2014-30.

7.   Le Coze, J.C., "New Models for New Times. An Anti-Dualist Move," *Saf. Sci.* 59 (2013): 200-218, doi:10.1016/j.ssci.2013.05.010.

8.   Malek, S., "What Software Developers Can Learn From the Latest Car Recalls," http://it-cisq.org/what-sofware-developers-can-learn-from-the-latest-car-recalls/, 2017.

9.   Altinger, H., Wotawa, F., and Schurius, M., "Testing Methods Used in the Automotive Industry: Results from a Survey," *Proc. of the 2014 Workshop on Joining AcadeMiA and Industry Contributions to Test Automation and Model-Based Testing-JAMAICA 2014*, 2014, ISBN:9781450329330, doi:10.1145/2631890.2631891.

10.   Whittaker, J.A., "What Is Software Testing? And why Is it So Hard?" *IEEE Softw.* 17, no. 1 (2000): 70-79, doi:10.1109/52.819971.

11.   Lions, J.L., "Ariane 5 Flight 501 Failure: Report by the Inquiry Board," 1996.

12.   Weyuker, E.J., "Testing Component-Based Software: A Cautionary Tale," *IEEE Softw.* 15, no. 5 (1998): 54-59, doi:10.1109/52.714817.

13.   NHTSA, "Investigation Report: PE 16-007 (MY2014-2016 Tesla Model S and Model X)," 2017.

14.   Leveson, N.G., *New Safety Technologies for the Automotive Industry*, 3rd ed. (Detroit: SAGE, 2006).

15.   ISO, *Road Vehicles-Functional Safety (ISO 26262)* (Sage, 2011).

16.   Yu, H., Lin, C.-W., and Kim, B., "Automotive Software Certification: Current Status and Challenges," *SAE Int. J. Passeng. Cars - Electron. Electr. Syst.* 9, no. 1 (2016):74-80, doi:10.4271/2016-01-0050.

17.   Khastgir, S., Birrell, S., Dhadyalla, G., Sivencrona, H. et al., "Towards Increased Reliability by Objectification of Hazard Analysis and Risk Assessment (HARA) of Automated Automotive Systems," *Saf. Sci.* 99 (2017): 166-177, doi:10.1016/j.ssci.2017.03.024.

18.   Charette, R.N., "This Car Runs on Code," *IEEE Spectr.* 46, no. 3 (2009).

19.   Khastgir, S., Birrell, S., Dhadyalla, G., and Jennings, P., *Identifying a Gap in Existing Validation Methodologies for Intelligent Automotive Systems: Introducing the 3xD Simulator*, 3rd ed. (Sage, 2015), ISBN:9781467372664, doi:10.1109/IVS.2015.7225758

20.   Kalra, N. and Paddock, S.M., "Driving to Safety: How Many Miles of Driving Would It Take to Demonstrate Autonomous Vehicle Reliability?," *Transp. Res. Part A Policy Pract.* 94 December 2016: 182-193, doi:10.1016/j.tra.2016.09.010.

21. Transport Systems Catapult, "Taxonomy of Scenarios for Automated Driving," 2017.

22. Koopman, P. and Wagner, M., "Challenges in Autonomous Vehicle Testing and Validation," *SAE Int. J. Transp. Saf.* 4, no. 1 (2016): 15-24, doi:10.4271/2016-01-0128.

23. Wachenfeld, W. and Winner, H., "The New Role of Road Testing for the Safety Validation of Automated Vehicles," *Automated Driving* (2017): 419-435, ISBN:978-3-319-31893-6, doi:10.1007/978-3-319-31895-0_17.

24. Campbell, R.L., "Will the Real Scenario Please Stand Up?," *ACM SIGCHI Bull.* 24, no. 2 (1992): 6-8, doi:10.1145/142386.1054872.

25. Cockburn, A. and Fowler, M., "Question Time! About Use Cases," *ACM SIGPLAN Not.* 33, no. 10 (1998): 226-229, doi:10.1145/286942.286960.

26. Cockburn, A., "Structuring Use Cases with Goals," *J. Object Oriented Program* 5 (1997): 1-16.

27. Flage, R. and Aven, T., "Emerging Risk-Conceptual Definition and a Relation to Black Swan Type of Events," *Reliab. Eng. Syst. Saf.* 144 (2015): 61-67, doi:10.1016/j.ress.2015.07.008

28. Ergai, A., Cohen, T., Sharp, J., Wiegmann, D. et al., "Assessment of the Human Factors Analysis and Classification System (HFACS): Intra-Rater and Inter-Rater Reliability," *Saf. Sci.* 82 (2016): 393-398, doi:10.1016/j.ssci.2015.09.028

29. Aven, T., "On the Meaning of a Black Swan in a Risk Context," *Saf. Sci.* 57 (2013): 44-51, 10.1016/j.ssci.2013.01.016.

30. Khastgir, S., Dhadyalla, G., Birrell, S., Redmond, S. et al., "Test Scenario Generation for Driving Simulators Using Constrained Randomization Technique," SAE Technical Paper 2017-01-1672, 2017, doi:10.4271/2017-01-1672

31. Griessnig, G. and Schnellbach, A., "Development of the 2nd Edition of the ISO 26262," Stolfa J., Stolfa S., O'Connor R., and Messnarz R., eds., *Systems, Software and Services Process Improvement: EuroSPI 2017: Communications in Computer and Information Science* (Cham: Springer, 2017), 535-546, doi:10.1007/978-3-319-64218-5.

32. Louise Barriball, K. and While, A., "Collecting Data Using a Semi-Structured Interview: A Discussion Paper," *J. Adv. Nurs.* 19, no. 2 (1994): 328-335, doi:10.1111/j.1365-2648.1994.tb01088.x.

33. Robson, C. and McCartan, K., *Real World Research: A Resource for Users of Social Research Methods in Applied Settings*, 4th ed. (Wiley, 2016), ISBN: 9781405182416.

34. Saldaña, J., *The Coding Manual for Qualitative Researchers*, 3rd ed. (Sage, 2016).

35. Robinson-Mallett, C.L., "An Approach on Integrating Models and Textual Specifications," *Proceedings of the 2nd IEEE International Workshop on Model Requirement Engineering MoDRE 2012*, 2012, 92-96, doi:10.1109/MoDRE.2012.6360079.

36. Robinson-Mallett, C., Grochtmann, M., Köhnlein, J., Wegener, J. et al., "Modelling Requirements to Support Testing of Product Lines," *ICSTW 2010-3rd International Conference on Software Testing, Verification and Validation Work*, 2010, 11-18, doi:10.1109/ICSTW.2010.65.

37. Alexander, I., "Misuse Cases: Use Cases with Hostile Intent," *IEEE Softw.* 20, no. 1 (2003): 58-66, doi:10.1109/MS.2003.1159030.

38. Stamatis, D.H., *Failure Mode and Effect Analysis: FMEA from Theory to Execution*, 2nd ed. (Milwaukee: ASQ Quality Press, 2003), ISBN: 9780521190817.

39. Hoffman, R.R., Lintern, G., and Eitelman, S., "The Janus Principle," *IEEE Intell. Syst.* 19, no. 2 (2004): 78-80, doi:10.1109/MIS.2004.1274915

40. Salmon, P.M., Cornelissen, M., and Trotter, M.J., "Systems-Based Accident Analysis Methods: A Comparison of Accimap, HFACS, and STAMP," *Saf. Sci.* 50, no. 4 (2012): 1158-1170, doi:10.1016/j.ssci.2011.11.009.

41. Chen, S.T., Wall, A., Davies, P., Yang, Z. et al., "A Human and Organisational Factors (HOFs) Analysis Method for Marine Casualties Using HFACS-Maritime Accidents (HFACS-MA)," *Saf. Sci.* 60 (2013): 105-114, doi:10.1016/j.ssci.2013.06.009.

42. Baysari, M.T., Caponecchia, C., McIntosh, A.S., and Wilson, J.R., "Classification of Errors Contributing to Rail Incidents and Accidents: A Comparison of Two Human Error Identification Techniques," *Saf. Sci.* 47, no. 7 (2009): 948-957, doi:10.1016/j.ssci.2008.09.012.

43. Wiegmann, D. and Shappell, S., "Applying the Human Factors Analysis and Classification System (HFACS) to the Analysis of Commercial Aviation Accident Data," *Proc. of the 11th International Symposium on Aviation Psychology*, Columbus, Ohio, 2001.

44. Lee, W.S., Grosh, D.L., Tillman, F.A., and Lie, C.H., "Fault Tree Analysis, Methods, and Applications - A Review," *IEEE Trans. Reliab.* R-34, no. 3 (1985): 194-203, 10.1109/TR.1985.5222114.

45. Reay, K.A. and Andrews, J.D., "A Fault Tree Analysis Strategy Using Binary Decision Diagrams," *Reliab. Eng. Syst. Saf.* 78, no. 1 (2002): 45-56, doi:10.1016/S0951-8320(02)00107-2.

46. Abimbola, M., Khan, F., and Khakzad, N., "Risk-Based Safety Analysis of Well Integrity Operations," *Saf. Sci.* 84 (2016): 149-160, doi:10.1016/j.ssci.2015.12.009.

47. Leveson, N.G., "Applying Systems Thinking to Analyze and Learn from Events," *Saf. Sci.* 49, no. 1 (2011): 55-64, doi:10.1016/j.ssci.2009.12.021.

48. Leveson, N., "A New Accident Model for Engineering Safer Systems," *Saf. Sci.* 42, no. 4 (2004): 237-270, doi:10.1016/S0925-7535(03)00047-X.

49. Leveson, N.G., *Engineering a Safer World* (The MIT Press, 2011), ISBN: 9780262016629.

50. Fleming, C.H., Spencer, M., Thomas, J., Leveson, N., and Wilkinson, C., "Safety Assurance in NextGen and Complex Transportation Systems," *Saf. Sci.* 55 (2013): 173-187, doi:10.1016/j.ssci.2012.12.005.

51. Khastgir, S., Birrell, S., Dhadyalla, G., and Jennings, P., "Calibrating Trust on Automation in Vehicles through Knowledge: Introducing the Concept of Informed Safety," *Transp. Res. Part C Emerg. Technol. (Under Review).*

Theory of Collision Avoidance Capability in Automated Driving Technologies

Toshiki Kindo
Toyota Motor Corporation

Bunyo Okumura
Toyota Research Institute-North America

This paper proposes a theory to analyze the collision avoidance capability of automated driving technologies. The theory gives answers to a fundamental question whether automated vehicles fall into extreme conditions at all rather than another question how a vehicle reacts under extreme conditions (is it as safe as driver?). The theory clarifies the following matters: There are two types of hazards to cause collisions, cognitive hazards and behavioral hazards. Cognitive hazards are handled by controlling the upper limit speed of the automated vehicle including when stopped. There are two methods for handling behavioral hazards, preparation and response. The response known well is the coping method activated when the hazard is detected in the dynamic (operational) level. The preparation is the coping method operating at all time in the semantic (tactical) level. The collision condition in the semantic level is as follows, a collision occurs when the paths of two vehicles have a crossing point and the two vehicles drive on the crossing point at same time. The condition can be formulated as collision avoidance equation. Solving the equation means that the automated vehicle has prepared for the behavioral hazard before the hazard occurs. It is concluded that a collision avoidance capability consists of not only a response capability that supports the accuracy of collision avoidance in extreme conditions in the dynamic

CITATION: Kindo, T. and Okumura, B., "Theory of Collision Avoidance Capability in Automated Driving Technologies," SAE Technical Paper 2018-01-0044, 2018, doi:10.4271/2018-01-0044.

level but also a preparation capability that supports the accuracy to avoid reaching those extreme conditions in the semantic level. The preparation capability can be evaluated through stability analysis of the automated vehicle behavior given by the temporal backward simulation from each extreme condition. A remaining problem is how determine the upper limit of the hazards growing speed to which the automated vehicles should react.

Introduction

Collision avoidance systems have been studied for over 20 years [1]. Concrete collision avoidance systems such as automatic emergency braking (AEB) system have been proposed to help avoid collisions with pedestrians [2]. Since the methods of these systems primarily involve vehicle control for avoiding collisions in extreme conditions, the main focus of these systems is on planning after a hazard is detected (i.e. "response'").

Recent works describe situations both after hazard detection ("response'") and before hazard detection ("preparation"). Risk potentials are used in various planning methods for vehicle safety [3, 4]. A method described in [3] provides risk quantification based on the collision velocity achieved by an AEB system for pedestrian safety. A method described in [4] provides both safe and human-like behavior by taking into account both trajectory-induction and risk-prevention potentials. These methods are capable of successfully avoiding or reducing the risk of collision, and the minimization of risk generates a "preparation" response to potential hazard.

There are methods for decision making in autonomous vehicle systems that incorporate both "response" and "preparation". Early autonomous driving systems used finite state machines (FSMs) [5] and several heuristic techniques [6] to solve decision making problems. More recent approaches have addressed decision making problems by using learning based methods such as machine learning [7] and reinforcement learning [8]. However, such methods still do not clearly differentiate decisions between "response" and "preparation".

In order to pursue an argument to its logical conclusion in practical use, a framework is proposed to describe operational design domain (ODD) and object and event detection and response (OEDR) of automated driving, and to present a theory of collision avoidance capability on the framework.

The proposed framework has a hierarchical structure that corresponds to varying levels of planning by drivers, such as route plans, lane selection, speed control, and so on.

Key points of the framework may be understood by considering a traffic situation at a highway junction consisting of a main road with multiple lanes and a single-lane approach road (Figure 1).

One vehicle (vehicle a) is driving on the main road and another (vehicle b) is driving on the approach road. In this case, vehicle a is able to select which lane to drive in.

If vehicle a selects the inside lane that has no junction point with the approach road, vehicle b will be able to move into the main road on the outside lane without negotiation and without requiring a collision avoidance effort.

If vehicle a selects the outside lane that connects with the approach road, there is a possibility of collision between vehicle a and vehicle b. In this case, the vehicle's' trajectories are classified into three groups. The first group corresponds to a set of collision-free cases in

FIGURE 1 A traffic situation with two vehicles at a highway junction that has multiple lanes on the main road and a single lane on the approach road.

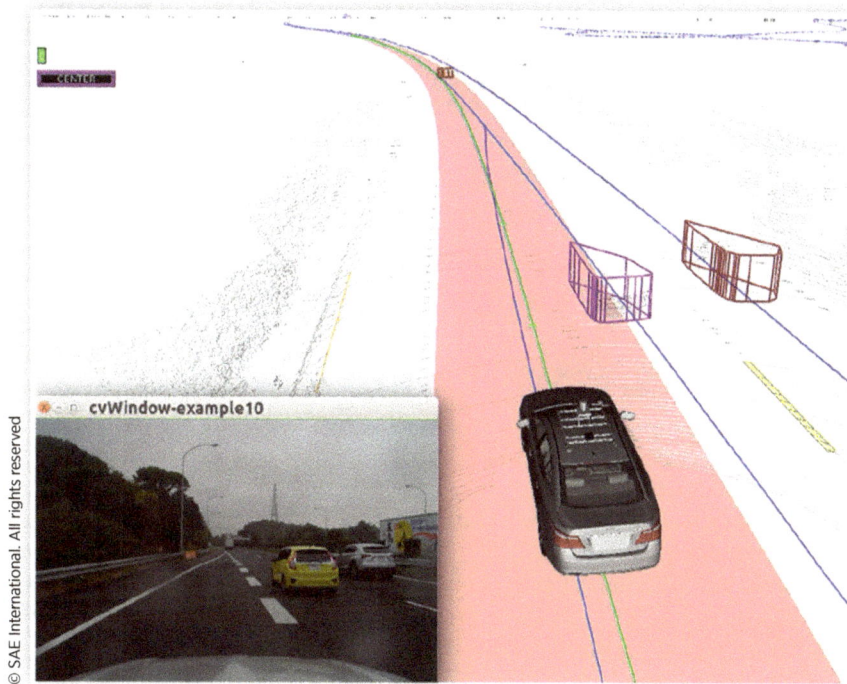

which vehicle *a* goes through the junction point before vehicle *b* reaches the same point. The second group corresponds to a set of cases in which the vehicles *a* and *b* collide at the junction point. The third group corresponds to a set of collision free-cases in which vehicle *a* goes through the junction point after vehicle *b* passed the same point. To avoid a collision, the vehicles have to select the first or third groups. An order-based decision making is required to ensure that vehicle *a* goes through the junction point. After the decision, each vehicle chooses and follows a better future trajectory from the trajectories limited by the decision.

The above case study shows that the decision making of each vehicle has a three-tiered process:

- lane selection,
- making an order-based decision to ensure that vehicle a goes through the junction point,
- choosing a better future trajectory.

The three-tiered process has the following properties:

- upper-level processing results affects to lower level processing results but lower results do not affect to upper results,
- the first and second levels of processing are semantic (tactical) decision making,
- the third level of processing is a best effort to generate dynamic (operational) behavior by the decision.

These properties strongly support the idea that considerable part of the collision avoidance capability of an automated vehicle can be described by the two upper levels of decision making.

Section 2 introduces a new definition of normal driving to support mathematical discussion in following sections. Section 3 provides several answers for ODD and OEDR listed by NHTSA's Federal Automated Vehicles Policy [9]. Section 4 formulates the decision making corresponding to the lane selection and the order-based decision making to ensure that vehicle *a* goes through the junction point. Section 5 discusses collision avoidance in "normal driving". Section 6 and 7 analyze the collision avoidance capability of automated vehicles in case when hazard occurs and in extreme traffic situations [10] with the proposed formulation.

Definition of Normal Driving

A description of normal driving based on various traffic scenes is given by NHTSA at "Federal Automated Vehicle Policy" [9, 11]. The scene based description has no clear dividing line between normal driving and otherwise. The normal driving description that is a set of vehicle behaviors in traffic scenes without collision is too complicate to conduct mathematical discussion on collision avoidance.

In this paper, a simple and general definition of normal driving is introduced to conduct mathematical discussion on collision avoidance. The new definition, "normal driving", refers to a situation in which the vehicle is directly driving in a lane at or under the speed limit defined by traffic rules. Any other situations are regarded as "hazard".

This definition of "normal driving" is very narrow compared with the above NHTSA's description. Precisely, it provides a basis, when vehicles with requisite sensing range discussed in the following section drive in "normal driving", all of the vehicles free from collision, for collision avoidance discussions.

Requisite Sensing Range

This section estimates the requisite sensing range to avoid a collision in "normal driving" using of a simple model of dynamics.

In "normal driving", one difficult collision-avoidance case involving deceleration is when automated vehicle a stops short of a motionless vehicle because the relative speed between automated vehicle a and the target vehicle at the maximum limit. To avoid a collision, the requisite sensing range must be longer than the braking distance with deceleration $a_a^{(-)}$

$$L^{(-)}(x) = \frac{1}{2}\frac{v_a^2(x)}{a_a^{(-)}}$$

where $v_a(x)$ is the speed of automated vehicle a with an upper speed limit at x provided by the Route Network definition File(RNDF).

Figure 2 shows another difficult case. Automated vehicle *a* is driving onto a priority road from a non-given priority road. This case is difficult because automated vehicle *a* needs much longer acceleration time compared with an intersection. This is a collision-avoidance case that requires positive acceleration. The actual acceleration time in these cases is 5 to10 seconds. To grantee that acceleration will not result in collision, the automated vehicle must detect no vehicle on the priority road that satisfies the following condition before entering the priority road.

$$x_b(0) - x_a(0)x_a(t) - x_a(0) + v_b t$$

FIGURE 2 Vehicle configuration in traffic situation in which automated vehicle *a* is entering a priority road from a non-priority road.

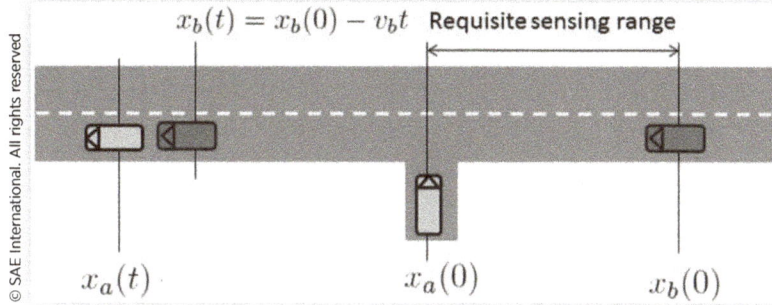

$$x_b(t) = x_b(0) - v_b t \quad \textbf{Requisite sensing range}$$

$$x_a(t) \qquad x_a(0) \qquad x_b(0)$$

where t is the acceleration time of vehicle a up to $v_a = v$, $x_a(0)$ is the initial position of vehicle a, $x_a(t)$ is the position of vehicle a at clock time t, $x_b(0)$ is the initial position of vehicle b and v_b is the speed of the vehicle b.

In "normal driving", v_b can be replaced with the speed limit of the priority road v and the acceleration time t with $v = a_a^{(+)}$ where $a_a^{(+)}$ is the standard acceleration rate of vehicle a. As a result, to avoid collision, the minimum distance between vehicles a and b at $t = 0$ is given by.

$$L^{(+)} = \frac{1}{2}\frac{v^2}{a_a^{(+)}}.$$

Thus, the requisite sensing range $L^{(+)}(x)$ for vehicle a is given by.

$$L^{(+)}(\boldsymbol{x}) = \frac{1}{2}\frac{v^2(\boldsymbol{x})}{a_a^{(+)}} + v(\boldsymbol{x})\Delta t_a.$$

where $v(x)$ is the speed limit at x provided by the RNDF, $a_a^{(+)}$ is the standard acceleration rate of vehicle a, and Δt_a is the effective time for automated vehicle a to turn.

Thus, the requisite sensing range is given by.

$$L_s^*(\boldsymbol{x}) = \max\left\{L^{(-)}(\boldsymbol{x}), L^{(+)}(\boldsymbol{x})\right\}.$$

Since the acceleration rate is generally smaller than the deceleration rate in "normal driving", the requisite sensing range satisfies.

$$L_s^*(\boldsymbol{x}) = L^{(+)}(\boldsymbol{x}).$$

in almost all cases

Operational Design Domain

From the previous discussion, it can be concluded that the operational design domain (ODD) in "normal driving" is given by.

$$\begin{aligned} L_s^\dagger \geq L_s^*(\boldsymbol{x}) &: \rightarrow x \in ODD, \\ L_s^\dagger L_s^*(\boldsymbol{x}) &: \rightarrow \boldsymbol{x} \notin ODD \end{aligned} \tag{1}$$

where $L_s^*(x)$ is the requisite sensing range at x and L_s is the effective sensing range of the automated vehicle. These conditions clarify the general cognitive requirements for OEDR as follows:

- an automated vehicle must understand the effective sensing range in which the vehicle detects all objects and events,
- when its effective sensing range is less than the requisite sensing range, an automated vehicle must change the upper limit of speed given by Equation (1) or send a "Take over request" to the driver.

The cognitive requirements are supported by

- evaluation of the localization accuracy,
- evaluation of the effective sensing range L_s^c,
- evaluation of the self-monitoring capability of effective sensing range.

These must be evaluated on test courses or defined standard routes on public roads.

Formulation of Collision Avoidance in the Semantic Level

Candidate Path Matrix

The discussion below makes the following assumptions:

- the automated vehicle's effective sensing range is larger than the requisite sensing range,
- each vehicle drives in a lane,
- vehicles are not driving side by side in a lane,
- the road network is represented by the RNDF, which is a simple graph consisting of dense way points.

For simplicity, traffic environment that include only vehicles is considered. It is easy to extend the discussion to include pedestrians and the like.

First, the candidate path matrix of vehicle a is defined as follows:

$$\mathcal{R}_{a\alpha} = \begin{cases} 1 & \text{the vehicle } a \text{ may drive at } \alpha, \\ 0 & \text{otherwise.} \end{cases}$$

where α is the index of the way points of the RNDF. We introduce interference matrix,

$$\mathcal{H}_{ab} = \sum_\alpha \sum_\beta \mathcal{R}_{a\alpha} \delta_{\alpha\beta} \mathcal{R}_{b\beta}.$$

where $\delta_{\alpha\beta}$ represents that two vehicles, a and b, are driving on the same way point. $\mathcal{H}_{ab} = 0$ means that the two vehicles a and b do not collide because there is no way point on which both vehicles are driving. $\mathcal{H}_{ab} \neq 0$ means that they have possibility of collision between the two vehicles, i.e., there is at least one way point where both vehicle a and b are driving. This way point is referred to as interference point x_{ab}^α.

COLLISION AVOIDANCE BY LANE SELECTION

This section considers a case where vehicle a is an automated vehicle and the interference matrix satisfies $\mathcal{H}_{ab} \neq 0$. If vehicle a follows the candidate path matrix $\mathcal{R}_{a\alpha}$, there is the possibility of collision with vehicle b.

One typical case, $\mathcal{H}_{ab} \neq 0$, is a case in which vehicle a is driving in the outside lane of a main road and vehicle b is driving on an approach road to the highway junction as discussed in Section 1. If vehicle a changes its driving lane from the outside to the inside lane, then the interference matrix becomes $\mathcal{H}_{ab} = 0$, which is equal to no possibility of collision. This is a successful case in which the change of the candidate path matrix $\mathcal{R}_{a\alpha}$ results in $\mathcal{H}_{ab} = 0$.

In contrast, when automated vehicle a is driving in the outside lane of the main road, it has no option to change its driving lane. In such $\mathcal{H}_{ab} \neq 0$ cases, automated vehicle a requires a scheduling effort for collision avoidance at the junction point.

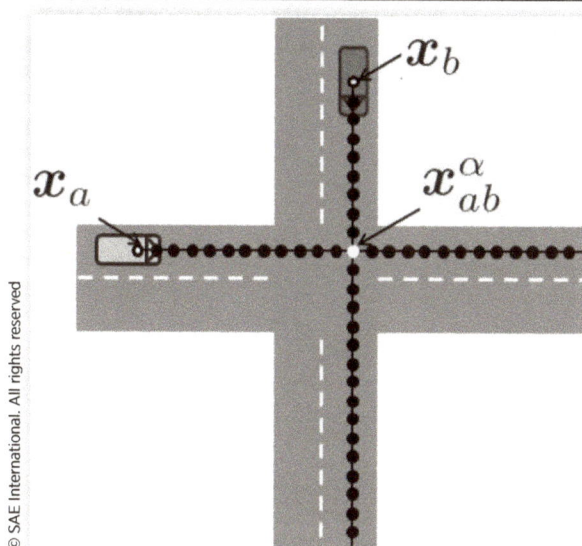

FIGURE 3 Traffic situation with two vehicles approaching a crossing.

TIMING TENSOR

The timing tensor is introduced here to represent scheduling in driving.

Consider the case shown in Figure 3.

The distance from vehicle a in position x_a to an arbitrary interference point is referred to as $d(x_a, x_{ab}^{\alpha})$. The traveling time from x_a to x_{ab}^{α} is given by.

$$t_{a\alpha} = \frac{d\left(x_a, x_{ab}^{\alpha}\right)}{v_a}$$

where v_a is the speed of vehicle a.

Similarly, the traveling time of vehicle b from x_b to x_{ab}^{α} is given by.

$$t_{a\alpha} = \frac{d(x_b, x_{ab}^{\alpha})}{v_b}$$

When $t_{a\alpha} = t_{b\alpha}$, vehicles a and b crash into each other at interference point x_{ab}^{α}. When $t_{a\alpha} < t_{b\alpha}$, vehicle a goes through interference point x_{ab}^{α} before vehicle b.

The timing tensor is as follows.

$$\mathcal{T}_{a\alpha,b\beta} = \begin{cases} \mathrm{sgn}\left(t_{b\alpha} - t_{a\alpha}\right)\delta_{\alpha\beta} & \text{at } x_{ab}^{\alpha}, \\ 0 & \text{otherwise.} \end{cases}$$

The sign of the timing tensor component $\mathcal{T}_{a\alpha,b\beta}$ shows the order of the vehicles passing the interference point. Accordingly, the timing tensor is antisymmetric for vehicle permutation $a \leftrightarrow b$,

$$\mathcal{T}_{a\alpha,b\beta} = -\mathcal{T}_{b\alpha,a\beta}.$$

Collision Avoidance Equation

To avoid a collision at interference point x_{ab}^{α}, the time offset δt between the times when vehicles a and b go through the point ($t_{a\alpha}$ and $t_{b\beta}$, respectively) must be sufficient.

To take account of the time offset, the timing tensor is revised as follows:

$$\mathcal{T}_{a\alpha,b\beta}(\delta t) = \begin{cases} \mathrm{sgn}\left(t_{b\beta} - t_{a\alpha} - \dfrac{1}{2}\delta t\right) & \text{, at } x_{ab}^{\alpha}, \\ 0 & \text{, otherwise.} \end{cases}$$

The revised timing tensor $\mathcal{T}_{a\alpha,b\beta}(\delta t)$ has both an antisymmetric part and a symmetric part for the vehicle permutations. The symmetric part of the revised timing tensor is given by.

$$\mathcal{C}_{a\alpha,b\beta}(\delta t) = \frac{1}{2}\left\{\mathcal{T}_{a\alpha,b\beta}(\delta t) + {}^{t}\mathcal{T}_{a\alpha,b\beta}(\delta t)\right\}$$

where $\mathcal{C}_{a\alpha,b\beta \neq a\alpha}(\delta t) = 0$. The non-zero component of $\mathcal{C}_{a\alpha,b\beta}(\delta t)$ shows that vehicle a and vehicle b collide at interference point x_{ab}^{α} when the temporal difference.

$$\Delta t = \left(t_{a\alpha} - t_{b\alpha}\right)$$

is less than δt. Accordingly $\mathcal{C}_{a\alpha,b\beta}(\delta t)$ is called the "conflict matrix".

To obtain the conflict matrix, only two types of observables are required, the position of the vehicle x, and the speed of the vehicle v. Thus, the conflict matrix including all possible collision cases is written as follows:

$$\mathcal{C}_{a\alpha,b\beta} = \mathcal{C}_{a\alpha,b\beta}\left(x_{a}, v_{a}, x_{b}, v_{b}, \delta t\right)$$

where time offset δt is the time margin parameter to avoid a collision.

Finally the collision possibility between vehicles a and b is given by the following collision matrix as an extension of the interference matrix

$$\begin{aligned} &\mathcal{D}_{ab}\left(x_{a}, v_{a}, x_{b}, v_{b}, \delta t\right) \\ &= \sum_{\alpha\beta} \mathcal{R}_{a\alpha}\left(x_{a}\right)\mathcal{C}_{a\alpha,b\beta}\left(x_{a}, v_{a}, x_{b}, v_{b}, \delta t\right) R_{b\beta}\left(x_{b}\right). \end{aligned}$$

Collision avoidance is accomplished when a solution is found that satisfies.

$$\mathcal{D}_{ab}(x_{a}, v_{a}, x_{b}, v_{b}, \delta t) = 0.$$

This is the proposed "collision avoidance equation" in the semantic level.

The following discussion abbreviates positions to x_{a} and the like for simplicity.

Collision Avoidance in Normal Driving

Unfortunately the existence of a solution for Equation (2) is not guaranteed in general. For example, if $\mathcal{R}_{b\beta}$ includes all way points that vehicle b can physically reach, there is no solution for the collision avoidance equation in the semantic level. Under loose constraints, vehicle b driving on a road without a center median has the possibility of reaching all way points in both the lanes extending in the same direction as the vehicles as well as in the opposing lanes. In this case an automated vehicle will never find a collision-free trajectory.

The fact shows that an automated vehicle needs constraints to evaluate other vehicle behavior. Considering that the constraints must be applicable to other vehicles driving on various types of roads, the constraints should be simple and general.

From this standpoint, the definition of "normal driving" introduced in the previous section is a reasonable constraint.

Collision Avoidance in Normal Driving

In order to avoid a collision with another vehicle, an automated vehicle must detect the position x_b and the speed v_b of all other vehicles b, $b = 1,2,\cdots$.

In "normal driving", the candidate path matrix of vehicle b, $\mathcal{R}_{b\beta}(x_b)$, is determined by its position x_b as follows: $\mathcal{R}_{b\beta}(x_b) = 1$ when the way point β is directly traced from position x_b and $\mathcal{R}_{b\beta}(x_b) = 0$ otherwise. The speed of vehicle b, v_b, is estimated by the speed limit $v(x_b)$ on the position x_b, $v_b = v(x_b)$ based on the traffic rules in the RNDF.

Consequently the collision avoidance equation in terms of $\mathcal{R}_{a\alpha}(x_a)$ and v_a is given by.

$$\mathcal{D}_{ab}\left(\boldsymbol{x}_a, v_a, \boldsymbol{x}_b, v_b(\boldsymbol{x}_b), \delta t\right) = 0. \tag{3}$$

Automated driving vehicle a has a two-stage response to solve the collision avoidance equation. The first response is a planning level response obtained through searching candidate path $\mathcal{R}_{a\alpha}(x_a)$ to minimize the value of $\mathcal{H}ab$. The second response is a scheduling level response obtained through searching speed v_a resulting in.

$$C_{a\alpha,b\beta}\left(\boldsymbol{x}_a, v_a, \boldsymbol{x}_b, v(\boldsymbol{x}_b), \delta t\right) = 0$$

with the planning level response

Preparation and Response to Hazard

There are two types of hazards, cognitive hazards and behavioral hazards. A typical cognitive hazard is when the sensing module fails to operate during driving. A typical behavioral hazard is when another vehicle appears to make a lane change. Each hazard generates the possibility of collision but does not lead directly to a collision. If an automated vehicle can make the appropriate preparation and response to each hazard that occurs, it can be avoid a collision.

Preparation and Response to Cognitive Hazard

An automated vehicle is not able to detect a cognitive hazard ahead of time but it constantly makes preparations for these hazards. One of the preparations is that an automated vehicle generates both a normal trajectory and a Fall Back trajectory that enables it to stop on the current lane without a collision. As a result, the automated vehicle can switch from the normal trajectory to the Fall Back trajectory when a cognitive hazard is detected.

If the sensing range of the automated vehicle is long enough, the Fall Back trajectory should be a connected two-part trajectory. The first part should be equivalent to the "normal driving" trajectory and the second part should be a trajectory that stops the

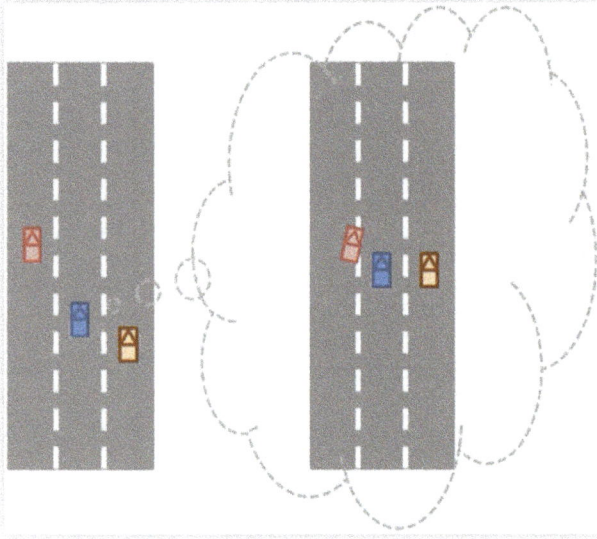

FIGURE 4 Typical behavioral hazard with lane change.

vehicles on the current lane without a collision. The first part gives time for the driver to take over the control of the automated vehicle smoothly.

Preparation and Response to Behavioral Hazard

Figure 4 shows a traffic situation of a three-lane road to clarify the key points of preparation and response to a behavior hazard. An automated vehicle is driving in the center lane. Two other vehicles are driving in the left and the right lanes, respectively. The automated vehicle is trying to pass by the vehicle driving in the left lane (the left vehicle).

Here the following notation of the candidate path matrix $\mathcal{R}_{a\alpha}$ is used with a defined as the index of vehicles and α as the index of the lanes and way points. The automated vehicle, the vehicle driving in the left lane, and the vehicle driving in the right lane are referred by $a = 1,2,3$, respectively α is decomposed into two indices as follows.

$$\alpha = (l,w)$$

where l and w are the lane index and the index of way points in the lane, respectively. The left lane, center lane, and right lane are referred to as $l = 1,2,3$.

In this case, the candidate path matrix of the left vehicle in "normal driving" is given by.

$$\mathcal{R}^n_{2(l,w)} = \begin{pmatrix} 0 & 0 & 1 & 1 & 1 & \cdots & 1 & 1 & 1 \\ 0 & 0 & 0 & 0 & 0 & \cdots & 0 & 0 & 0 \\ 0 & 0 & 0 & 0 & 0 & \cdots & 0 & 0 & 0 \end{pmatrix}$$

where the row and the column represent the lane and the way points on the lane, respectively. A solution of the collision avoidance equation in "normal driving".

$$\sum_{\alpha\beta} \mathcal{R}_{1\alpha} C_{1\alpha,2\beta} \mathcal{R}^n_{2\beta} = 0$$

is

$$\mathcal{R}^0_{1(l,w)} = \begin{pmatrix} 0 & 0 & 0 & 0 & 0 & \cdots & 0 & 0 & 0 \\ 1 & 1 & 1 & 1 & 1 & \cdots & 1 & 1 & 1 \\ 0 & 0 & 0 & 0 & 0 & \cdots & 0 & 0 & 0 \end{pmatrix}$$

as the candidate path matrix of the automated vehicle. A natural speed profile of the solution is provided by a constant function, $so(w) = v_1$.

Considering a behavioral hazard in which the left vehicle start to change lanes from the left to the center lane as the automated vehicle catches up with the left vehicle, the "hazard" candidate path matrix of the left vehicle is given by

$$\mathcal{R}^h_{2(l,w)} = \begin{pmatrix} 0 & 0 & 0 & 0 & 0 & \cdots & 0 & 0 & 0 \\ 0 & 0 & 0 & 1 & 1 & \cdots & 1 & 1 & 1 \\ 0 & 0 & 0 & 0 & 0 & \cdots & 0 & 0 & 0 \end{pmatrix}$$

and the collision avoidance equation taking account of the behavioral hazard is.

$$\sum_{\alpha\beta} \mathcal{R}_{1\alpha} \mathcal{C}_{1\alpha,2\beta} \left(\mathcal{R}_{2\beta}^{n} + \mathcal{R}_{2\beta}^{h} \right) = 0.$$

This equation has two types of solutions, the "vigilant solution" and the "efficient solution".

In the vigilant solution, the candidate path matrix is the same as $\mathcal{R}_0^{a\alpha}$. Here, the automated vehicle drives in the center lane behind the left vehicle. The driving efficiency of this solution is less than the driving efficiency of the "normal driving" because the speed of automated vehicle is reduced to the speed of the left vehicle. The longitudinal position of the vigilant solution is different from the position of the normal driving solution without the bifurcation point at which the normal driving solution and the vigilant solution coincide. As a result, the automated vehicle is not able to switch its driving plan from the vigilant solution to the normal driving solution after selecting the vigilant solution.

In the efficient solution, the speed profile $s^{1st}(w)$ is the same as $so(w)$. This means that the driving efficiency of the solution equals the efficiency of "normal driving". In this case, the candidate path matrix of the solution $\mathcal{R}_{1(l,w)}^{1st}$ corresponds to the automated vehicle changing lanes from the center to the right lane. According to this solution, the automated vehicle changes lanes from the center to the right lane in response to the left vehicle lane change and overtakes the left vehicle. Furthermore the automated vehicle is able to switch its driving plan between the efficient solution and the normal driving solution at any time because the efficient solution has same longitudinal position as the solution for "normal driving".

The switching capability of the driving plan enables the automated vehicle to responds each hazard independently. If the automated vehicle takes account of a possible additional collision with the right vehicle, then the collision avoidance equation changes from a single equation to the following simultaneous equation.

$$\sum_{\alpha\beta} \mathcal{R}_{1\alpha} \mathcal{C}_{1\alpha,2\beta} (\mathcal{R}_{2\beta}^{n} + \mathcal{R}_{2\beta}^{h}) = 0,$$

$$\sum_{\alpha\beta} \mathcal{R}_{1\alpha} \mathcal{C}_{1\alpha,3\beta} \mathcal{R}_{3\beta}^{n} = 0.$$

The above efficient solution satisfies the first equation of the simultaneous equation but does not satisfy the second equation. In contrast, the normal driving solution does not satisfy the first equation of the simultaneous equation but satisfies the second equation. Therefore the automated vehicle is able to avoid a collision by switching the driving plan from the normal driving solution to the efficient solution without the synchronous case in which the automated vehicle catches up with the left vehicle at the same timing as the right vehicle catches up with the automated vehicle.

Next, this section considers the situation from the viewpoint of the right vehicle instead of the automated vehicle. From this new viewpoint, the switching of the auto-mated vehicle driving plan, which is a natural response of the automated vehicle to the hazard generated by the left vehicle, presents a new hazard for the right vehicle. This suggests that easily switching the driving plan may cause a hazard avalanche in traffic situations.

Consequently an automated vehicle has to consider the effect of switching driving plans on other vehicles. If the effect requires one of other vehicles to react, the driving plan switching should be suppressed. Typically an automated vehicle must select the vigilant solution until it has been passed by the right vehicle in the synchronous case when it has to respond to the left and right vehicles at the same time.

In a similar manner, the collision avoidance equation with N vehicles is represented by a simultaneous equation consisting of N equations as follows,

$$\sum_{\alpha\beta} \mathcal{R}_{1\alpha} \mathcal{C}_{1\alpha,b\beta} \mathcal{R}_{b\beta} = 0$$

where $b = 2, 3, \cdots, (N+1)$. Each equation is expanded around "normal driving" as follows.

$$\sum_{\alpha\beta} \mathcal{R}_{1\alpha} \mathcal{C}_{1\alpha,b\beta} \left(\mathcal{R}_{b\beta}^{n} + \mathcal{R}_{b\beta}^{h} \right) = 0.$$

When a number of hazards is introduced:

$$N^{h} = \sum_{b=1}^{N} \rho_b \left(N^h \right)$$

where $\rho_b(N^h) = 1$ when vehicle b is the vehicle that deviates from "normal driving" and $\rho_b(N^h) = 0$ otherwise, an expanded simultaneous equation around "normal driving" is obtained as series of hazards numbering N^h.

$$\sum_{\alpha\beta} \mathcal{R}_{1\alpha} \mathcal{C}_{1\alpha,b\beta} \left(\mathcal{R}_{b\beta}^{n} + \rho_b \left(N^h \right) \mathcal{R}_{b\beta}^{h} \right) = 0 \tag{4}$$

where $b = 2, 3, \cdots, (N+1)$ and $Nh = 0, 1, 2, \cdots$.

The expanded simultaneous Equation (4) is classified by N^h as follows

- a "normal driving" simultaneous equation with $N^h = 0$, its solution is an efficient solution when the automated vehicle believes that all vehicles drive in "normal driving",

- a single hazard simultaneous equation with $N^h = 1$, its solution is a vigilant solution to take account of N single hazards,

- a double hazard simultaneous equation with $N^h = 2$, its solution is a very vigilant solution considering $N(N-1)/2$ combinations of two hazards,

- •••.

At the first step, the collision avoidance equation referred to as the "normal driving" equation $N^h = 0$ is solved to obtain the "normal driving" solution (0th solution) and the speed profile of the efficient solution. At the second step, that is, the preparing step for single hazards, a solution for the single hazard equation is found with the same speed profile of the 0th solution. However, if there is no solution with the same speed profile, then another solution with a new speed profile should be identified. The solution obtained through the second step (1st solution) is an efficient or a vigilant driving plan.

When the hazard possibility dissipates, the equation that should be solved becomes the "normal driving" equation again. If the solution is an efficient driving plan, the efficient driving plan is smoothly switched to the new "normal driving" plan because their speed profiles are same. On the other hand, if the solution is a vigilant driving plan, there is a transitional state because there is difference between the speed profile of the vigilant driving plan and that of the new "normal driving" plan. Then the driving plan slowly converges on the new "normal driving" plan through the transitional state.

From a mathematical viewpoint, it is possible to obtain 2nd, 3rd, and higher solutions of the expanded simultaneous equation. However the practical values of the 2nd or higher solutions must be assessed carefully for the following reasons. The first reason is the expected properties of the higher solutions. An increase in the number of cases increases the number of interference points. Accordingly, it will become harder to find a driving plan with the same speed profile as the "normal driving" plan. As a result, the solution is likely to converge on a very vigilant one in which the automated vehicle drives in its current lane at low speed. The second reason is the amount of calculation. When the automated vehicle counts of N^h hazards, the number of cases considered is $_N C_N{}^h$, which is huge. The third reason is that the event probability of multiple hazards is, however, quite low compared with that of a single hazard.

In this context, the expansion of the simultaneous equation around "normal driving" based on the number of hazards is a perturbation expansion based on the event probability. Therefore, a problem is on what terms the hazard number expansion of the collision avoidance equation should be truncated. The practical truncating number of hazards is expected to be 1 because the automated vehicle selects the vigilant plan when 2 or more hazards are detected.

This section can be summarized as follows:

1. an automated vehicle makes "normal driving" and switchable driving plans to avoid various single hazards as collision avoidance preparation,
2. the automated vehicle drives following the "normal driving" plan,
3. when the automated vehicle detects an actual hazard, it switches from the "normal driving" plan to the prepared plan corresponding to the detected hazard as the collision avoidance response.

Finally, we would like to point out that there is a remaining problem. It is obvious that an automated vehicle is not able to avoid a collision with a hazard whose growing speed is faster than the switching speed from the "normal driving" plan to the efficient driving plan. Such rapidly growing hazards include attack of maliciously controlled vehicles discussed in [12]. Accordingly, if the automated vehicle is required to react to rapidly growing hazards with a very small event probability, the automated vehicle will always select the vigilant driving plan. How to determine the upper limit of the hazards growing speed to which the automated vehicle should react is a remaining problem for achieving the practical application of these vehicles.

In addition, the results of this section show that various properties of a vehicle are outlined by a normal behavior model referred to as "normal driving" and the hazard growing speed on the collision avoidance problem. Accordingly, the collision avoidance problem with a pedestrian and the like will be solved using suitable behavior model in the same manner.

Behavior Analysis in Extreme Traffic Situations/Hazards

Understanding how a vehicle reacts under extreme conditions (is it as safe as a human driver?) is an important question [10]. There are, however, more fundamental questions that should be asked, such as whether automated vehicles fall into extreme conditions at all shown at Figure 5. This is because the previous section made it clear that automated vehicles with a high

FIGURE 5 A schematic figure represents posterior behavior (solid line) and anterior behavior (dashed line) in an extreme condition. In order to remove"?" that is a question - whether automated vehicles fall into extreme conditions at all, numerical simulations in the temporal backward direction from the extreme condition is required.

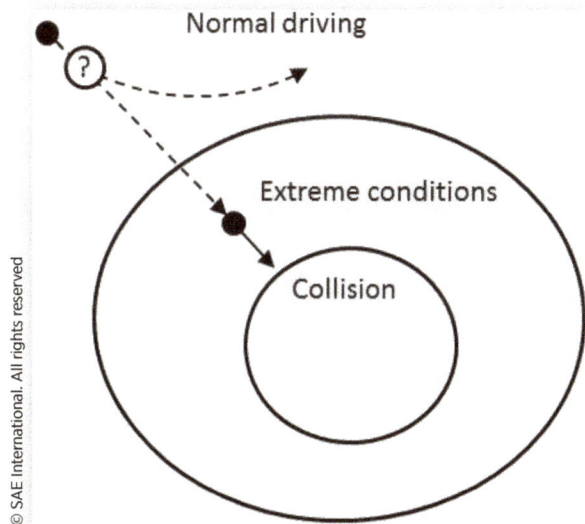

preparation capability are able to stay off the extreme conditions. Therefore, when considering the behavior of an automated vehicle in extreme conditions, not only the posterior behavior of the automated vehicle after the extreme condition but also the anterior behavior before the extreme condition should be analyzed. Analysis of anterior behavior requires demands numerical simulations in the temporal backward direction from the extreme condition.

Stability of Extreme Condition

The time evolution equation of vehicle a is given by.

$$\xi_a\left(t+\delta_T\right)=T_a\left(\delta_T\right)\xi_a\left(t\right)$$

where ξ_a is the dynamic state of vehicle a, $T_a\left(\delta\tau\right)$ is the time evolution operator for vehicle a. The time evolution of the traffic environment with N vehicles is given by.

$$E\left(t+\delta_T\right)=T\left(\delta_T\right)E\left(t\right)$$

where $E(t) = t(\xi_1(t), \xi_2(t), \cdots, \xi_N(t))$ and a time evolution operator.

$$T\left(\delta_T\right)=\begin{cases} T_1\left(\delta_T\right) & 0 & \cdots & 0 \\ 0 & T_2\left(\delta_T\right) & & 0 \\ \vdots & & \ddots & 0 \\ 0 & 0 & \cdots & T_N\left(\delta_T\right) \end{cases}.$$

Classical mechanics states that, if the sequence of the traffic environment in an extreme condition $E_e(t)$.

$$E\left(t-n\delta t\right)\rightarrow\cdots\rightarrow E\left(t-\delta t\right)\rightarrow E^e\left(t\right). \qquad (5)$$

is unstable, then the event probability of extreme condition $E_e(t)$ is zero. This means that the automated vehicle does not fall into extreme condition $E_e(t)$. Such an extreme condition is referred to as an "unstable extreme condition".

If all extreme conditions in a set of extreme conditions including an automated vehicle $\{E_e\}$ are unstable, it can be concluded that the automated vehicle will not collide in the extreme conditions without analyzing the automated vehicle response to the extreme conditions.

However, if stable extreme conditions are present, the automated vehicle response must be analyzed. That is, how does the vehicle react under extreme conditions (is it as safe as a human driver?) .

Trace-Back of Traffic Environment and Stability

This section outlines a method to analyze automated vehicle behavior before extreme conditions occur.

It is difficult to derive the trace-back operator $T^{-1}\left(\delta\tau\right)$ to get $\xi(t - \delta\tau)$ from $\xi(t)$ analytically. Here, an iterative method is adopted to obtain the operator.

The minimum description of the dynamic state of the vehicle a, includes position x_a, velocity v_a, heading ϕ_a, and angle speed of heading ω_a as follows:

$$\xi_a = \begin{Bmatrix} x_a \\ v_a \\ \phi_a \\ \omega_a \end{Bmatrix}.$$

Then, $\xi_a(t - \delta\tau)$ is given by.

$$\xi_a(t-\delta\tau) = \xi_a(t) - \delta\tau \frac{d}{dt}\xi(t) + o(\delta t^2)$$

$$\sim \left[1 - \delta\tau \begin{Bmatrix} 0 & 1 & 0 & 0 \\ 0 & 0 & 0 & 0 \\ 0 & 0 & 0 & 1 \\ 0 & 0 & 0 & 0 \end{Bmatrix} \right] \xi(t)$$

$$= T^{-1}(\delta\tau)\xi(t)$$

where $T^{-1}(\delta\tau)$ is the linear-trace back operator.

When the vehicles in the traffic environment have mutual interaction, the linear trace-back operator $T^{-1}(\delta\tau)$ is not equal to the inverse operator of the time evolution operator $T(\delta\tau)$.

$$e(t) \neq E(t)$$

where.

$$e(t) = T(\delta_T)T^{-1}(\delta_T)E(t).$$

Now the following iterative equation can be considered.

$$e_{i+1}(t) = T(\delta\tau)T^{-1}(\delta\tau)E_i(t)$$
$$\Delta E_{i+1}(t) = e_{i+1}(t) - E_0(t)$$
$$E_{i+1}(t) = E_i(t) - \gamma\Delta E_{i+1}(t)$$

where γ is a small positive free parameter to control the convergence speed and.

$$E_0 = E^e(t)$$

The above $e_{i+1}(t)$ converges on $Ee(t - \delta\tau)$ satisfying.

$$E^e(t) = T(\delta_T)E^e(t - \delta_T).$$

Thus, the temporal sequence (5) can be obtained by applying the method iteratively.

The stability of the temporal sequence (5) is evaluated by the following analysis: A tiny variation is introduced around $Ee(t)$,

$$E_p^e(t - n\delta_T) = E^e(t - n\delta_T) + \Delta E_p(n),$$

where $p = 1, 2, \cdots, K$ and K is the dimension number of the dynamic state of the automated vehicle. The tiny variation is transformed into by the time evolution. The relationship between the tiny variations and the transformation result are represented as.

$$T\left(\Delta E_p(n)\right) = T(\delta\tau)E_p^e(t - n\delta\tau) - T(\delta\tau)E^e(t - n\delta\tau)$$

$$T\left(\Delta E_p(n)\right) = \Lambda(n)\Delta E_p(n)$$

where $\Lambda(n)$ is $K \times K$ matrix and its eigenvalues are.

$$\lambda_1(n) \geq \lambda_2(n) \geq \cdots \geq \lambda_K(n).$$

If $1 \geq \lambda_1(n)$, then the temporal sequence (5) is stable at step n. If $1 < \lambda_1(n)$, then it is unstable. The stability of the sequence is evaluated by $\prod_n \lambda_1(n)$.

Summary of Forward and Backward Analysis

Temporal forward analysis of extreme conditions clarifies the response capability of the automated vehicle based on the final result, i.e., whether a collision occurs. In contrast, temporal backward analysis of extreme conditions clarifies the preparation capability of the automated vehicle based on the stability of the temporal sequences of the traffic environment, i.e., whether the automated vehicle enters an extreme condition. As a result, Table 1 is obtained. It is important that the "Collision" and "Unstable" section of Table 1 is labeled "No collision" because the automated vehicle does not enter the extreme condition.

Discussion

The collision area of Table 1 can be broken down into several sections based on "normal driving" and hazard situations as shown in Table 2. The first "+" or "-" sign in the "normality" sections indicates the driving status of the automated vehicle and the second "+" or "-" sign indicates the driving status of the other vehicles." +" indicates "normal driving" and "-" indicates otherwise.

TABLE 1 Summary table of forward/backward analysis

		Backward analysis(Stability)	
		Unstable	Stable
Forward	No Collision	No Collision	
analysis(Collision)	Collision	Collision	Collision

TABLE 2 Substructure of collision area of behavior analysis table

			Stable			
			Anterior normality			
Collision			–	–+	+–	++
Collision	Posterior normality	–	Avoidable by T/O	–+	+–	
		TBD			++	
			NG			

FIGURE 6 A typical unavoidable case. The light gray vehicle is driving straight on at the speed limit. The darker vehicle suddenly turns right at higher speed than the speed limit suddenly.

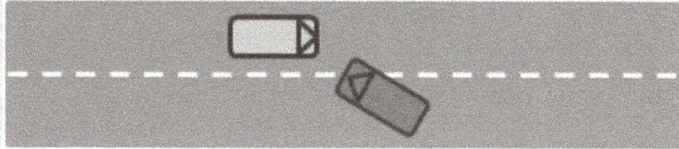

When the status of the automated vehicle is not "normal driving", the automated vehicle is able to detect abnormalities in its own status, by comparing its driving states to the traffic rules of the RNDF. The section referred to as "avoidable by T/O" in Table 2 indicates the section in which the automated vehicle is able to send the driver a take-over(T/O) request before a collision by identifying an abnormality.

The section referred to as "NG" contrasts with the "avoidable by T/O" section. In this section, the status of both the automated vehicle and other vehicles are "normal driving". Therefore both the automated vehicle and the other vehicles expect that the number of collision cases to be zero.

The remaining section of the collision area is referred to as "TBD". In this section, the automated vehicle drives following the "normal driving" plan and the other vehicles generate hazards. Since there are unavoidable cases in actual situations, the number of cases classified into this section is not zero. One of the typical unavoidable cases is the very rapid turn of an oncoming vehicle shown in Figure 6. In this case, the automated vehicle carries out full braking as the best effort to avoid a collision, even if it is not able to do so. The final result of this case is determined by the balance between the stopping ability - one of the maneuver abilities - of the automated vehicle and the rapidity of the oncoming vehicle - the hazard growing speed. In general, because the upper limit of the rapidity is higher than that of the stopping ability due to the properties of the vehicles, the number of collisions is not zero. The actual number depends on the balance described above.

This theoretical conclusion of behavior analysis in extreme conditions is the exactly same as the conclusion of preparation and response to a hazard, which is how to determine the upper limit of the hazard growing speed to which an automated vehicle should react. How to balance the upper limit of the hazard growing speed and the maneuver abilities of the automated vehicle is a remaining problem (the balance problem) for achieving the practical application of automated vehicles.

The balance problem should be solved as a social problem rather than a technical problem because an automated vehicle cannot control the hazard growing speed.

Summary/Conclusion

A theory to analyze the collision avoidance capability of automated vehicles is proposed. The structure of the theory is summarized at Figure 7. The theory clarifies the importance of the preparation, which is the collision avoidance method operating at all time in the semantic (tactical) level and which avoids for the automated vehicle to fall into extreme traffic environments as a precursor state before the collision. There is a remaining problem referred to as the balance problem, how balance the upper limit of the hazard growing speed and the maneuver abilities of the automated vehicle for achieving the practical application of automated vehicles.

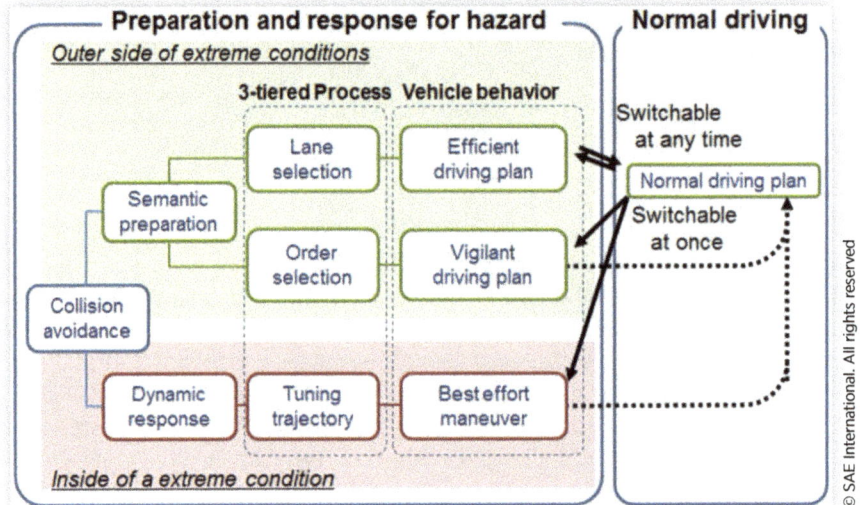

Contact Information

Dr. Toshiki Kindo
TOYOTA MOTOR CORPORATION
Higshifuji Technical Center
1200, Mishuku, Susono, Shizuoka, 410-1193 Japan
toshiki_kindo@mail.toyota.co.jp

Acknowledgments

The authors would like to express their gratitude to Masahiro Harada and Naoki Nagasaka from TOYOTA RESEARCH INSTITUTE-NORTH AMERICA (TRI- NA) for their inputs into discussions about the problems of automated vehicles passing through a roundabout and so on. The problems extracted from field tests using TRI-NA 's automated test vehicles and the subsequent discussions provided the impetus to start this research.

References

1. Lemelson, J. and Pedersen, R., "GPS Vehicle Collision Avoidance Warning and Control System and Method," US Patent 5,983,161, 1999.

2. Coelingh, E., Eidehall, A., and Bengtsson, M., "Collision Warning with Full Auto Brake and Pedestrian Detection-a Practical Example of Automatic Emergency Braking," *Intelligent Transportation Systems (ITSC), 2010 13th International IEEE Conference on*, 2010, 155-160.

3. Shimizu, T. and Raksincharoensak, P., "Motion Planning via Optimization of Risk Quantified by Collision Velocity Accompanied with AEB Activation," *Vehicular Electronics and Safety (ICVES), 2017 IEEE International Conference on*, 2017, 19-25.

4. Guo, C., Kidono, K., Machida, T., Terashima, R. et al., "Human-Like Behavior Generation for Intelligent Vehicles in Urban Environment Based on a Hybrid Potential Map," *2017 IEEE Intelligent Vehicles Symposium (IV)*, 2017, 197-203.

5. Montemerlo, M., Becker, J., Bhat, S. et al., "Junior: The Stanford Entry in the Urban Challenge," *Journal of Field Robotics* 25, no. 9 (2008): 569-597.

6. Urmson, C., Anhalt, J., Bagnell, D. et al., "Autonomous Driving in Urban Environments: Boss and the Urban Challenge," *Journal of Field Robotics* 25, no. 8 (2008): 425-466.

7. Okumura, B., James, M., Kanzawa, Y. et al., "Challenges in Perception and Decision Making for Intelligent Automotive Vehicles: A Case Study," *IEEE Transactions on Intelligent Vehicles* 1, no. 1 (2016): 20-32.

8. Galceran, E., Cunningham, A., Eustice, R., et al., "Multipolicy Decision-Making for Autonomous Driving via Changepoint-Based Behavior Prediction," *Robotics: Science and Systems* (2015).

9. The U.S. Department of Transportation, "Federal Automated Vehicles Policy, September 2016," https://www.transportation.gov/AV/federalautomated-vehicles-policy-september-2016.

10. OICA, "OICA Position on Automated Vehicle Certification under Type Approval," *Informal Working Group on Intelligent Transport Systems-Automated Driving ITS/AD-12 -11*, 2017.

11. Nowakowski, C., Shladover, S., Chan, C. et al., "Development of California Regulations to Govern the Testing and Operation of Automated Driving Systems," *California PATH Program*, Berkeley, University of California, November 14, 2014, at 10, http://docs.trb.org/prp/15-2269.pdf.

12. Dadras, S., Gerdes, R., and Sharma, R., "Vehicular Platooning in an Adversarial Environment," *Proceedings of the 10th ACM Symposium on Information, Computer and Communications Security*, 2015, 167-178.

A Lane-Changing Decision-Making Method for Intelligent Vehicle Based on Acceleration Field

Bing Zhu, Shuai Liu, and Jian Zhao

Jilin University

Taking full advantage of available traffic environment information, making control decisions, and then planning trajectory systematically under structured roads conditions is a critical part of intelligent vehicle. In this paper, a lane-changing decision-making method for intelligent vehicle is proposed based on acceleration field. Firstly, an acceleration field related to relative velocity and relative distance was built based on the analysis of braking process, and acceleration was taken as an indicator of safety evaluation. Then, a lane-changing decision method was set up with acceleration field while considering driver's habits, traffic efficiency and safety. Furthermore, velocity regulation was also introduced in the lane-changing decision method to make it more flexible. Afterwards, the polynomial trajectory planning method was matched up with this lane-changing decision-making method and simulations based on Matlab/Simulink were finally conducted to verify the method presented in this paper. As the simulation results showed, adopting the lane-changing decision-making method based on acceleration field, the lane-changing measurements such as starting position, span and driving speed can be optimized with driver's habits involved. At the same time, the vehicle safety can be well ensured.

CITATION: Zhu, B., Liu, S., and Zhao, J., "A Lane-Changing Decision-Making Method for Intelligent Vehicle Based on Acceleration Field," SAE Technical Paper 2018-01-0599, 2018, doi:10.4271/2018-01-0599.

Introduction

Recently, intelligent vehicle technology has widely attracted attention in both military and civilian fields, and some entrepreneurs have begun researching and testing [1]. However, most intelligent vehicles can only travel under limited conditions, and there are still many issues need to be solved. Among these problems, lane-changing decision-making is a critical section for an intelligent vehicle to perform as well as or even better than a human driver. Defined as a part of intelligent vehicle system's upper control module, lane-changing decision-making and subsequent trajectory planning can significantly affect traffic efficiency, driving safety and ride comfort.

In order to determine the factors that influence a driver's decision on lane-changing, Gipps.P.G. proposed the Gipps model, which assumes that the driver's behavior is rational, and decision-making depends on safety, feasibility, obstacles' location and advantage in relative speed in target lane [2]. MITSIM model and the CORSIM model are used more frequently. In MITSIM model, the process of a lane change is divided into three steps: judgement for the necessity of a lane change, detection of the gap between traffic vehicles, and implementation of a lane change. This model puts forward two concepts: tolerance factor and speed difference factor, which are used to judge the necessity of a change; once given the target speed, traffic signal and vehicle position information, the model will judge feasibility; when both necessity and feasibility satisfy requirements, a lane-changing behavior will be executed [3]. The CORSIM model makes decisions by motivation, benefits and urgency [4]. Jongsang Seo designed a decision-making algorithm, which consists of vehicle location index module, mode selection module and direction selection module. The vehicle location index module is responsible for receiving external environment information, the mode selection module is used to judge whether a collision will happen and the direction selection module is used to select the proper target lane [5]. These approaches offer much reference on lane-changing decision-making method. However, they consider only few elements involved. Drivers' habits, uncertainty of traffic vehicles' motion, and road adhesion condition can all have effects on the method. To make further exploration, Xuemei Chen used rough set theory to extract lane-changing rules from data obtained from a driving simulator. She divided the lane-changing process into two phases: intention generation phase and implementation phase. Analysis results demonstrate that during intention generation phase, decisions almost depend on only two factors: relative distance and relative velocity between host vehicle and preceding vehicle. During the implementation phase, if relative velocity is large, then relative distance has little effect on the decision; when relative distance is small, relative velocity would impose a great effect on the decision [6]. Jieyun Ding came up with the notion of Comprehensive Decision Index, which was designed with fuzzy method to estimate surrounding traffic vehicles' effect on decision-making [7]. Ding Zhao proposed a rapid evaluation method for intelligent vehicle, which was based on data collected under cut-in scenarios [8]. These methods built on pattern recognition require a large amount of experimental data to fit different driving habits. But without a model with clear physical meaning, some of those outputs can hardly be predictable and their accuracy needs to be further improved.

In fact, a complete lane-changing process involves trajectory planning as well, which entails an optimal start position and an initial velocity. Traditional artificial potential field methods are suitable for decision-making and trajectory planning under a static environment, but they comes with defects of falling into the local minimum and can lead to jittering when a vehicle is close to an obstacle [9], which makes them unable to adapt to a complex traffic environment. One way to skip the trap of local minimum in trajectory planning is to use rapid-exploring random tree algorithm, or RRT algorithm. It was proposed by LaValle S. M. to search for an accessible path [10]. The method is characterized by simplicity and efficiency. In order to overcome certain disadvantages of the algorithm itself [11,12], Qingkun Jiang proposed an improved RRT algorithm with B spline curve to make the planned trajectory smooth and executable [13]; Song J Z put forward an improved algorithm of bidirectional extend RRT, which was designed for practical application on intelligent vehicles [14]; adopting a goal-biased sampling strategy and a reasonable metric function, Mingbo Du presented a continuous-curvature RRT algorithm to reduce the waste of computing resource and accelerate planning process [15].

Although artificial potential field method has the advantage of continuity and the RRT method has the advantage of efficiency, neither of them is suitable for trajectory planning under urban environment, since the former is bound to produce a shaking trajectory and the latter lacks stability. After comparison and analysis, this paper adopts the polynomial trajectory planning method to match the proposed decision making method and ensure trajectory curve's continuity at the same time.

In this paper, we proposed a lane-changing decision-making method. Compared with other methods mentioned above, this method is more concise, and physical meaning is clear. Besides, the result is predictable, which is crucial for intelligent vehicles. Its innovation lies in the introduction of continuous acceleration field to provide guidance for personalized lane-changing decision-makings. Taking various factors into consideration, the decision-making method can be more practical and reliable.

Remaining parts are arranged as follows: In Section II, based on analysis of braking process, an artificial acceleration field related to relative velocity and relative distance is set up. Section III takes advantage of the field and proposes a lane-changing decision-making method. In section IV, velocity regulation is taken into consideration. In section V, the lane-changing decision-making method and the polynomial trajectory planning method are linked together, and simulations based on Matlab/Simulink are conducted to verify the superiority of the method.

Acceleration Field

Generally, two periods will be experienced successively within the process of confirming a lane-changing decision: intention generation and feasibility judgement. Based on existing research, we can find out factors that prompt a driver to change lane, such as relative velocity and relative distance. These parameters both have something to do with braking safety. Therefore, in order to predict lane-changing intention quantitatively, we extend emergency braking safety distance model first, and then establish an acceleration field.

Braking Safety Distance Model

Braking safety distance refers to the distance between host vehicle and the traffic vehicle ahead from which time the host vehicle start to brake and can finally maintain the same speed as preceding vehicle without collision. In braking safety distance model, assumptions are as follows:

1. Host vehicle and preceding vehicle are in the same urban straight lane.
2. Preceding vehicle keeps driving at constant but lower speed than host vehicle's initial speed.
3. Time for braking reaction will not be taken into consideration.

FIGURE 1 Acceleration variation curve in a braking process.

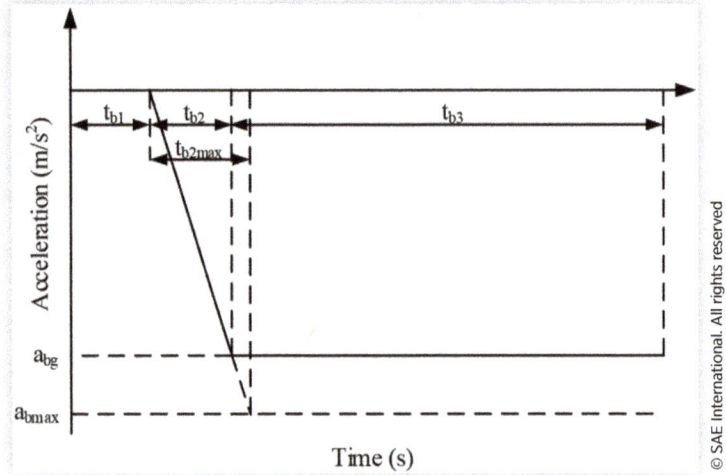

Then host vehicle's braking process can be divided into three stages: elimination of brake clearance, rise in braking pressure, and constant deceleration braking. The curve of acceleration-time is shown in Figure 1. Note that there is no necessity to keep on braking when velocity of host vehicle is reduced to the same value as front vehicle, so the braking process ends at that moment.

t_{b1} in the figure represents the time for eliminating brake clearance, during which time the acceleration is zero, and distance traveled by the host vehicle can be expressed as:

$$S_{b1} = v_h t_{b1} \tag{1}$$

Where v_h is the velocity of host vehicle.

t_{b2} in the figure represents the time for increasing braking pressure. In this period, acceleration is reducing with the slope of k, and the distance traveled by host vehicle can be expressed with S_{b2}:

$$k = \frac{a_{b\,max}}{t_{b2\,max}} \tag{2}$$

$$S_{b2} = \int\left[v_h + \int\left[a(t) \right] dt \right] dt = \frac{v_h}{k} a_{bg} + \frac{a_{bg}^3}{6k^2} \tag{3}$$

Where $a_{b\mathrm{max}}$ is the minimum acceleration accessible, which is negative, a_{bg} is the target acceleration during the braking process, which can be set as any value within $a_{b\mathrm{max}}$ at will.

t_{b3} in the figure represents the time for uniform acceleration braking. During t_{b3}, braking deceleration remains constant until the host vehicle's speed is down to the same value as the front vehicle's. The distance traveled by the host vehicle can be expressed as:

$$S_{b3} = \frac{v_t^2 - v_h^2}{2a_{bg}} - \frac{a_{bg}^3}{8k^2} - \frac{v_h a_{bg}}{2k} \tag{4}$$

Where v_t is the velocity of preceding vehicle.

During the whole braking process, the traveling distance of preceding vehicle is

$$S_{bt} = v_t \left(t_{b1} + \frac{a_{bg}}{2k} + \frac{v_t - v_h}{a_{bg}} \right) \tag{5}$$

FIGURE 2   Position relationship in the braking process.

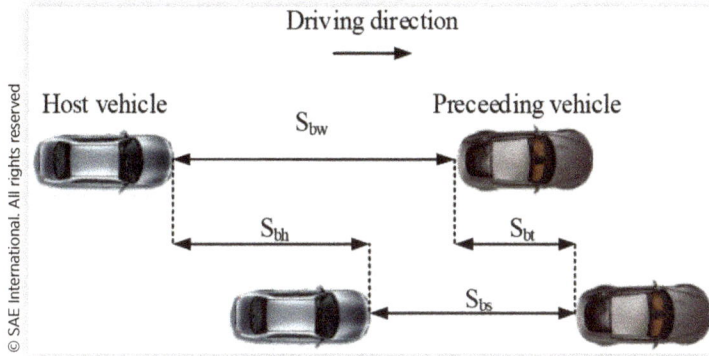

In order to ensure safety, a certain margin S_{bs} should be kept between these two vehicles in the final state, as shown in Figure 2. And the braking safety distance can be deducted as in formula (6):

$$S_{bw} = S_{bh} - S_{bt} + S_{bs} = -\frac{\Delta^2 v}{2a_{bg}} - \Delta v\left(t_{b1} + \frac{a_{bg}}{2k}\right) + \frac{a_{bg}^3}{24k^2} + S_{bs} \tag{6}$$

s. t.

$$S_{bh} = S_{b1} + S_{b2} + S_{b3} \tag{7}$$

$$\Delta v = v_t - v_h \tag{8}$$

Given a target deceleration, the formula (6) can calculate the corresponding safety distance according to the current speed difference. The relationship between deceleration, speed difference and safety distance can also be expressed as in Figure 3.

FIGURE 3   Improved model of emergency braking safety distance.

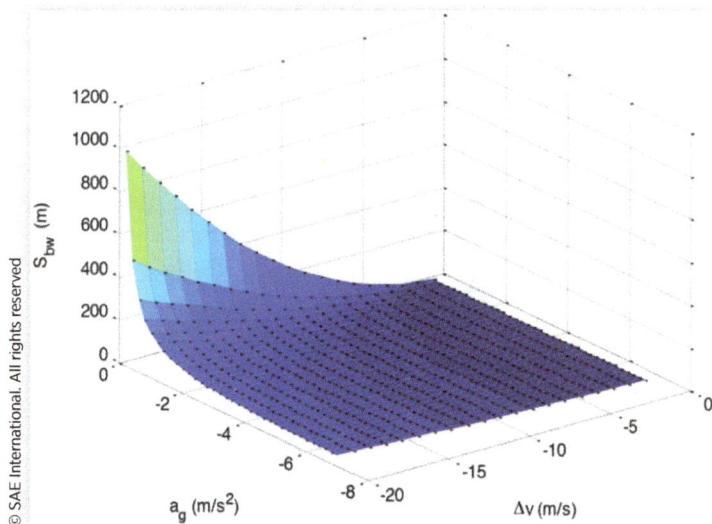

Acceleration Field for Intention Generating

In an emergency braking condition, target braking deceleration is expected to be the maximum that can be reached under a certain road condition. In fact, if we adjust the target braking acceleration to a different value, we can acquire a different braking safety distance, which can be regarded as an emergency braking safety distance model under a specific road condition, or as a non-emergency braking safety distance model on a better road. Thus, we are inspired to take different target braking accelerations to represent degrees of safety on a satisfactory pavement and as a parameter that drives the driver to generate the intention of making a lane change.

In formula (6), if we take S_{bw} and Δv as independent variables, a_{bg} as dependent variable, then we can derive the value of a_{bg} as S_{bw} and Δv change within the range, thus an acceleration field is formed, as in Figure 4.

Generally, when preceding vehicle on the same lane runs faster than host vehicle does, a driver can hardly be in want of a lane change. And when preceding vehicle's speed is lower than that of the host vehicle, with relative distance continue to reduce along with the influence of velocity difference, there is an increasing likelihood for a driver to produce the intention to change lane. Now we use S_{bw} to denote the current relative distance between host vehicle and preceding vehicle, Δv to denote the velocity difference, then substitute them into the numerical relationship, the calculation result a_{bg} can represent magnitude of a driver's desire to switch lane: stronger desire comes with greater value of $|a_{bg}|$. Note that parameter a_{bg} now indicates the target braking acceleration to be applied if the braking is taken at the moment and then achieve a critical state where collision happens to be avoided. This method of measuring a driver's degree of willingness to change lane is consistent with life experience from the view of ensuring safety, reducing speed loss, along with improving traffic efficiency.

FIGURE 4 Braking acceleration field.

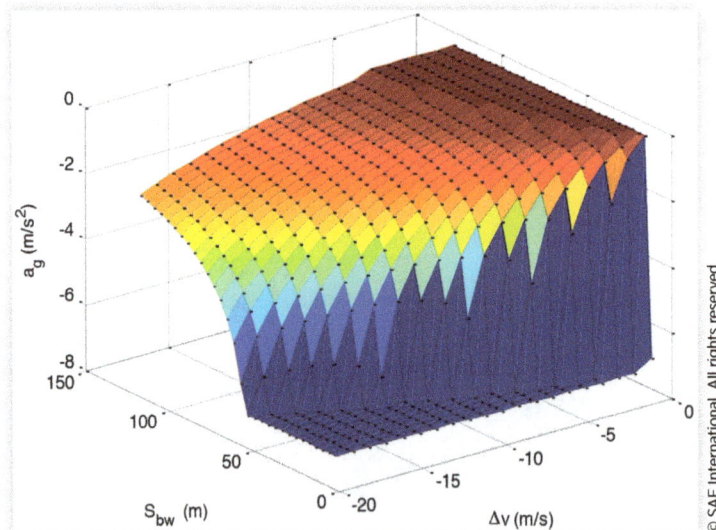

Acceleration Field for Feasibility Judgement

As mentioned above, when the way forward is 'blocked' by the vehicle ahead, a driver will naturally think of a lane change. Thus the second step of confirming a lane-changing decision needs to be performed, which aims at determining whether a lane change is feasible.

For front vehicle on the side lane, we can take it as on the same lane with the host vehicle and then compute a_{bg} by relationship expressed in formula (6), which is adopted as the indicator of lane-changing feasibility.

For rear vehicle on the side lane, if it runs slower than the host vehicle and keeps a certain distance away, a lane change will be feasible; if it runs faster than the host vehicle, the feasibility of a lane change will require a further judgment. In the lane-changing process, it is of higher priority for the passing of vehicle on the adjacent lane, and it is reasonable to assume the velocity of rear vehicle on the side lane as constant. In order to avoid collision, the host vehicle should be affected like an electrically charged particle when another particle with same electrical property comes near, and the force of repulsion can result in a forward acceleration motion. Therefore, we can derive the driving safety distance model with reference to the braking safety distance model, then substitute current relative distance and relative velocity into the model, the acceleration calculated in reverse can be used to quantitatively describe the strength of this effect.

Similarly, we analyze the process of driving and get the driving safety distance model as expression (9).

$$S_{dw} = S_{dt} - S_{dh} + S_{ds} = \frac{\Delta^2 v}{2a_{dg}} + \Delta v\left(t_{d1} + \frac{a_{dg}}{2k_d}\right) - \frac{a_{dg}^3}{24k_d^2} + S_{bs} \tag{9}$$

Where S_{dw} is the driving safety distance, S_{dt} is the distance traveled by rear vehicle during the host vehicle's accelerating process, S_{dh} is the distance traveled by the host vehicle in the process, S_{ds} is the margin that should be kept in the end of the process, a_{dg} is the target acceleration during the driving process, t_{d1} is the time for eliminating acceleration clearance, k_d is the growth rate of acceleration.

Complete Acceleration Field

Combining the braking and driving conditions, we can obtain the whole acceleration field as in Figure 5. A positive S_w stands for the relative distance between preceding vehicle and host vehicle, and a negative S_w represents the relative distance between the rear vehicle and host vehicle. When a traffic vehicle is running at a higher speed than host vehicle, Δv is a negative number. Otherwise, it is zero or a positive number. Figure 5 (a) is the 3-D diagram of the acceleration field, which covers the entire domain of definitions. Figure 5 (b) is the top view of (a), it clearly shows that dangerous situations are all embodied in the second and fourth quadrants: the blue part of the upper left corner indicates the need for braking, and the red part of the lower right corner calls for an acceleration.

Now, by calculating the acceleration a_g needed for not to collide with preceding vehicle on the same lane, we can roughly predict the likelihood of generating a lane-changing intention. Then, by calculating the acceleration needed for front side traffic vehicle and rear side traffic vehicle respectively, we can make further studies so as to judge the feasibility of that lane change.

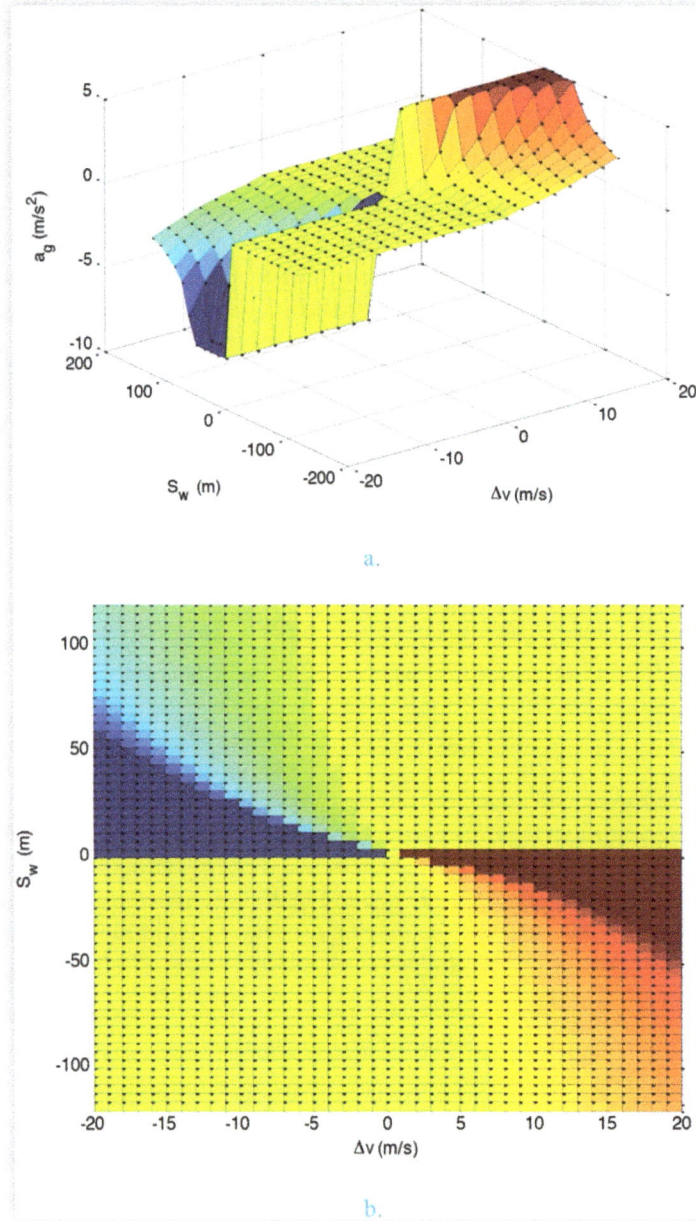

Basic Lane-Changing Decision-Making Method

A narrow gap, along with a velocity larger than that of preceding vehicle, can prompt the host vehicle to slow down or change lane. However, before deciding whether to make a lane change or not, the feasibility of a lane change will keep unknown until a refined calculation is conducted, so it's necessary to define a critical state, where worse situations can be regarded as dangerous and thus unacceptable. In doing so, we build a braking decision-making model and take the deceleration acquired as the safety base line.

Braking Decision-Making Method

One thing to reaffirm is that the acceleration (a negative number) obtained in the acceleration field represents the least braking intensity needed to ensure safety. Therefore this acceleration field can be taken as the basic safety model. It means that before a lane change is confirmed feasible, host vehicle is supposed to drive at a deceleration no less than the value calculated in the field.

On this foundation, we have established two decision reference lines and a warning line in a relative distance-relative velocity state plane, as in Figure 6.

Line B1: primary lane-changing intention reference line. It is a contour in the braking acceleration field. We assume the driver of host vehicle is rational, and the existence of lane-changing intention depends on safety index, which can be quantified in acceleration field. In this case, it's reasonable to set a specific reference acceleration value and extract the corresponding line B1 from the field. When acceleration calculated with actual traffic environment information reach this reference line, a human driver is highly likely to start to plan a lane change. It should be noted that the reference acceleration value is personalized and can be adjust to adapt to different types of drivers.

Line B2: brake starting reference line. This is also a contour of a certain acceleration value in the braking acceleration field. According to driving experience, in a relatively safe condition, a driver will usually not brake to reach the same speed as preceding vehicle, in which process there will be a loss of traffic efficiency. Besides, it does nothing good to improve ride comfort. So we set an acceleration threshold, when required braking intensity reaches the corresponding line B2, the driver would begin to consider braking. Similarly, this acceleration threshold can be adjusted to be personalized as well.

Line B3: emergency braking warning line. This line is calculated with the maximum braking deceleration the road surface can provide. When the state point in the plane corresponds to actual traffic environment is below the line B3, the host vehicle can only brake with the maximum deceleration accessible, even if a collision is inevitable.

With these definitions, a braking decisions can now be illustrated with Figure 6. Suppose there is a vehicle in front of the lane, and the relative speed and relative distance between preceding vehicle and host vehicle can be represented by point A on the state plane. Since we have already assumed the velocity of the traffic vehicle is invariant, if host vehicle does not take any action, velocity difference between them will stay the

FIGURE 6 Braking decision-making reference lines.

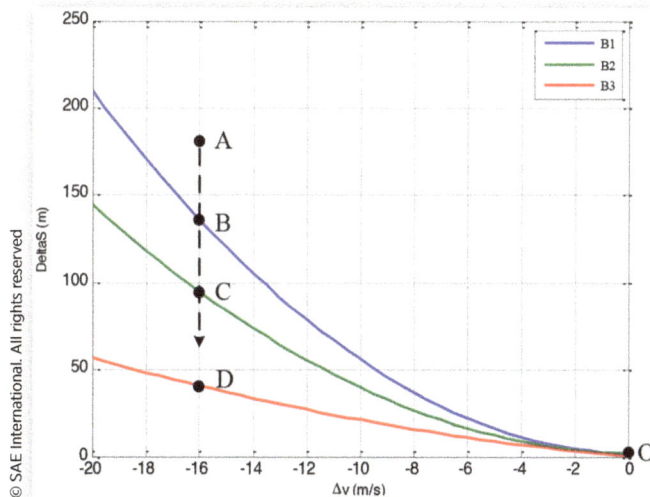

same, but relative distance will gradually decrease over time. During this process the state point will move from point A to point B, C and D successively, and the deceleration required by the field will increase. Therefore we suggest the host vehicle not to take any action when state point is above the line B1, but to make judgements all the time after reaching that point, which means the driver may have already sensed the potential danger. And then do as follows:

1. If driving risk can be reduced by an available lane change, then make it immediately (it will be elaborated in detail later);

2. If it's unfeasible to make a lane change at the time and the state point is on the line segment BC, since it has not reached brake starting line B2, there is not necessity to brake. In this case, the host vehicle should keep on driving without taking extra operation;

3. If it's unfeasible to make a lane change at the time and the state point is below point C, the host vehicle is expected to brake at the deceleration calculated in the field to ensure safety.

For extreme conditions corresponding to state point D or below, if a lane change is still unpractical, a collision is bound to occur and what host vehicle can do is to brake at maximum deceleration to reduce loss as much as possible.

Basic Lane-Changing Decision-Making Method

LANE-CHANGING ASSUMPTIONS AND SIMPLIFICATIONS

After setting up a braking decision-making method that guarantees safety, we need to judge the feasibility of a potential lane change when there has already been a lane-changing intention. Therefore, we make reasonable assumptions and simplifications of a lane-changing process:

1. Speeds of both host and traffic vehicles remain unchanged;

2. To make it conservative, host vehicle is seen as in both starting lane and target lane during the process;

3. On the premise of ensuring safety, a lane change is supposed to give priority to velocity loss. Therefore, only when velocity of front side vehicle is lower than that of host vehicle, but larger than that of preceding vehicle on the same lane with host vehicle, in which process the host vehicle has to finally keep the same speed as the vehicle in front of it, a lane change can be considered beneficial to efficiency and then carried out.

4. As a simplified case, we will temporarily not consider host vehicle's speed regulation before a lane change, which means the host vehicle can only adjust to the velocity of preceding vehicle on the target lane after the completion of that change.

ANALYSIS OF LANE-CHANGING PROCESS

In the process of planning, the system should generate a series of trajectories [16], which include an optimal trajectory and a worst one within the limit of tolerance. Assumed invariant during the lane-changing process, for each driving velocity, there is an optimal and a threshold lane-changing duration. According to these two time points, we obtain the corresponding distance required to complete the lane change under different relative velocities, as shown in reference line LC1 and line LC2 in Figure 7.

FIGURE 7 Lane-changing decision-making reference lines.

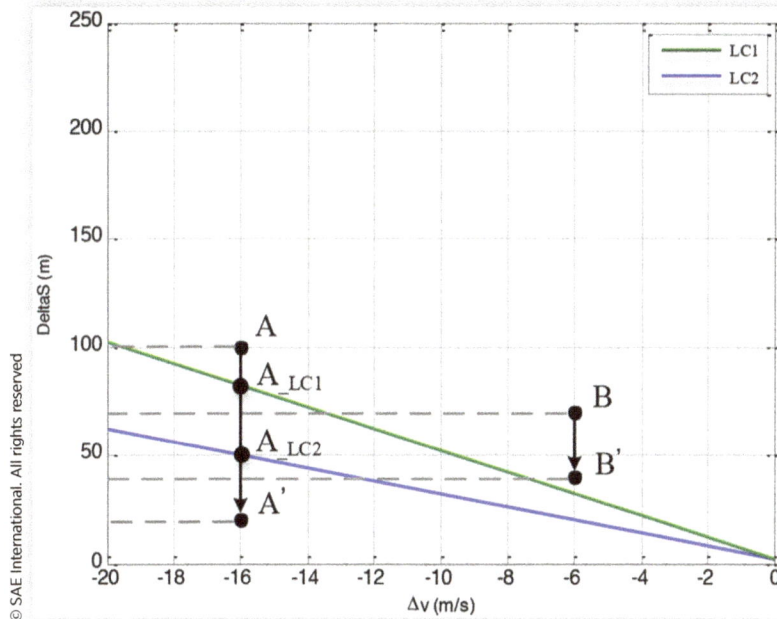

Line LC1: optimum lane-changing reference line. When state point is on or above line LC1, host vehicle can take the optimal time length to change lane. After getting line LC1, we take the larger value between line LC1 and the primary lane-changing intention reference line B1 (in Figure 6) to form a new lane-changing intention reference line, so as to ensure enough space for a lane change when intention is generated.

Line LC2: lane-changing emergency warning line. When the state point is between line LC1 and line LC2, duration of an expected lane change should be shorter than the optimum value and longer than the lower limit, and it can be proportionally derived from the state point's position on the plane. When state point is below line LC2 and host vehicle is assumed to be driving at a constant speed during the lane-changing process, a lane change can no longer avoid conflict with front car, then braking is the only way to mitigate collision loss.

A typical lane-changing process can be shown in Figure 8. When host vehicle has the idea of changing lane, it will have to make judgement on feasibility. We use semi-transparent icons in the graph to indicate relative position between vehicles at the time. At the beginning of decision-making, we denote the relative distance between host vehicle and traffic vehicle 1 as S_A, and distance between host vehicle and traffic vehicle 2 as S_B. They are marked by point A and point B on the state plane respectively.

FIGURE 8 Relative distance relationships in a typical lane-changing process.

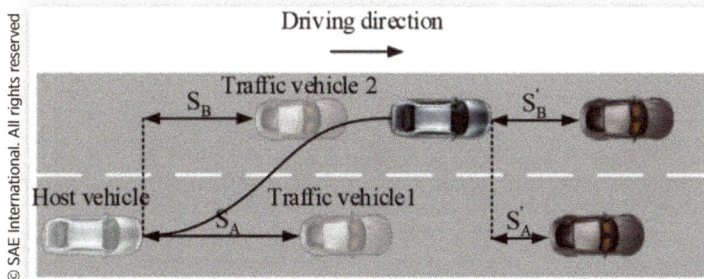

As the point A is above point A$_{LC1}$ on line LC1, host vehicle is supposed to change lane with the optimum lane-changing time. Then relative distances after the completion of a lane change can be recorded by S'_A and S'_B, which are corresponding to point A' and B'. Since we assume the host vehicle is on both original and target lane for the duration, the process can be reflected by state trajectories on state plane as line AA' and BB'.

For line AA', the final sate point A' is above the line where $DeltaS = 0$, which means the relative distance is larger than 0, thus safety condition meet our requirement. For line BB', there's necessity to calculate the deceleration needed on the final state point B' in acceleration field to estimate the risk from traffic vehicle 2. If the brake intensity calculated is lower than a permissible value, the process can be finally evaluated as safe and feasible. Otherwise, the plan of a lane change at that moment should be cancelled.

DECISION-MAKING METHOD

According to analysis of braking and lane-changing process, we can summarize a lane-changing decision-making method as in Figure 9.

FIGURE 9   Decision-making flow.

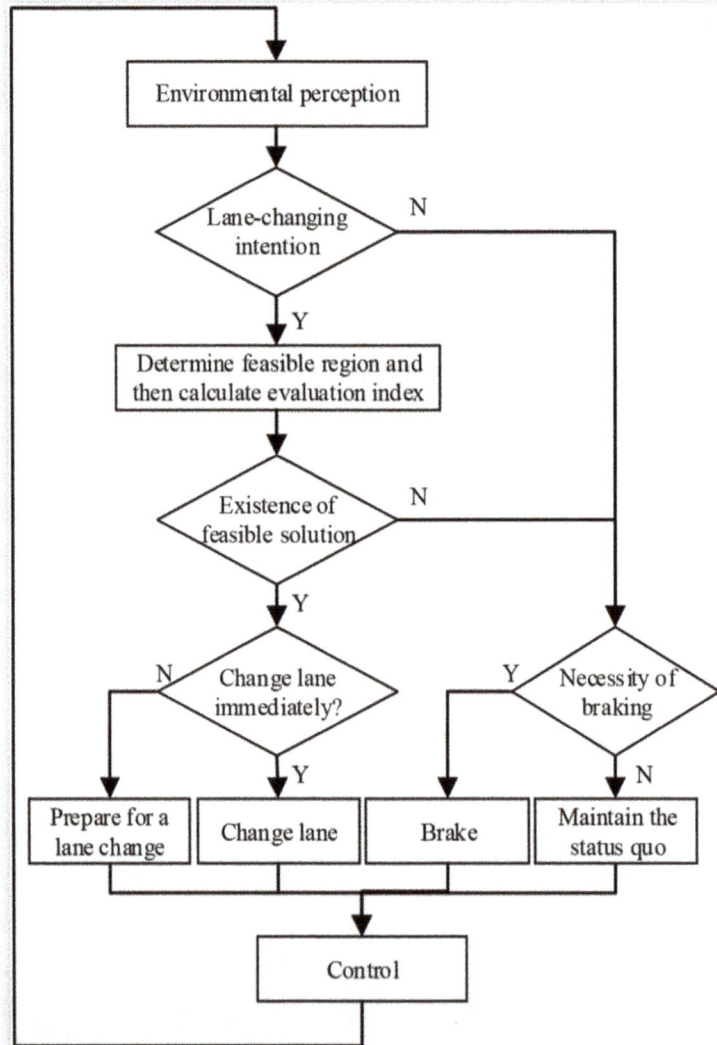

In each decision-making cycle, host vehicle should first use on-board sensors to perceive surrounding environment information, including relative distance and relative velocity between host vehicle and traffic vehicles nearby.

Then determine if there is a lane-changing intention: without an intention, it should drive in a straight line as instructed by the braking decision-making method; otherwise, it's necessary to judge the feasibility of a lane change by lane-changing decision-making method. As simplified in this section, where we ignore velocity regulation, then choose "brake", "change lane" or "maintain the status quo" directly. When practicable, change lane at once, otherwise, host vehicle is supposed to run in the braking mode within current decision-making cycle.

Extended Lane-Changing Decision-Making Method

In this section, we will extend the basic lane-changing decision-making method in two aspects:

1. Consider the impact from rear side traffic vehicle;
2. Consider velocity regulation before it starts to change lane.

Inspection on Rear Side Traffic Vehicle

In the basic lane-changing decision-making method, we have considered potential risks from preceding vehicles. Moreover, it's meaningful to judge the feasibility from rear side traffic vehicle's perspective.

According to traffic rules and driving experience, a lane change usually has no effect on the vehicle ahead. And a vehicle can only change lane when poses no danger to rear vehicle on the side lane. For rear vehicle, it has to adjust speed to adapt to real-time dynamic traffic environment. This process involves interaction between two vehicles, thus we take two parameters to judge the feasibility of a possible lane change. The first one is the acceleration acquired from the acceleration field for host vehicle, which is exerted by rear vehicle on the side lane. The second one is the deceleration acquired for rear vehicle, which assumes that host vehicle will keep a constant speed. These two parameters represent the urgency imposed on two vehicles separately, and any of them exceeding an acceptable range will reject the claim for a lane change.

Extended Method with Velocity Regulation

In Section II, we allow the braking process to be executed only after the completion of a lane change. In fact, host vehicle can certainly brake before changing lane, which will gain flexibility at the expense of part of safety. The process of velocity regulation can be illustrated with Figure 10. Relationship with preceding vehicle and anterolateral vehicle can be represented by point A and B respectively. When planning a lane change, host vehicle needs to determine the feasible region for velocity regulation.

Since the host vehicle will eventually run at the same speed with front vehicle in the target lane, and speed difference between host vehicle and the two vehicles ahead during the process of deceleration is reduced synchronously, so we stipulate that the maximum range of velocity regulation for host vehicle equals the initial speed difference between the host vehicle and the target front vehicle. Besides, to make it safe, starting position of

FIGURE 10 Sketch map of lane-changing decision-making method with velocity regulation.

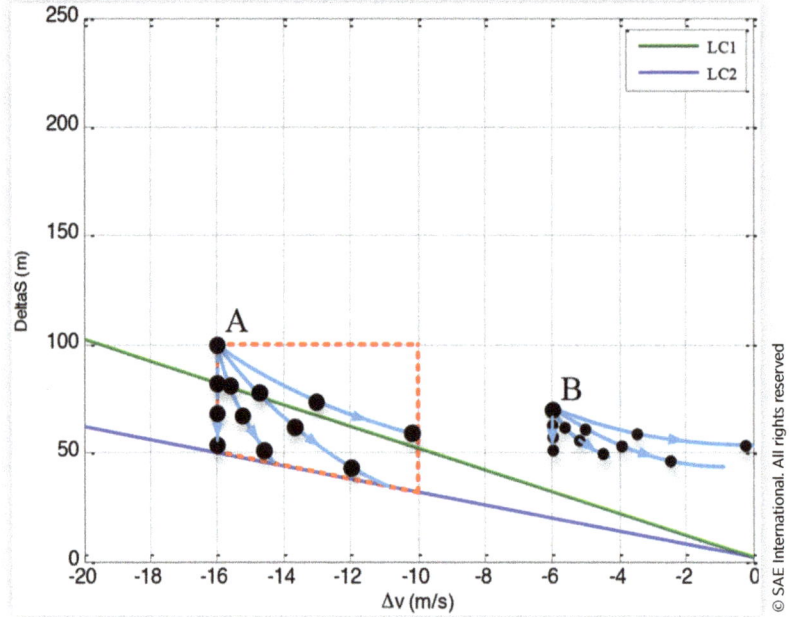

a lane change on state plane shall not be lower than the lane-changing emergency warning line. Then we can determine a feasible regulation region, which consists of a speed regulation range and a distance regulation range, shown as the red frame in Figure 10.

In attempt to find the optimal initial state of lane change, we select a series of acceptable acceleration and then calculate the state points at a certain time interval, which are represented by black dots and taken as possible initial states of a lane change. Next, screen out those feasible with the method presented in Section II. After that, make a comprehensive evaluation of each plan by function designed as expression (10).

$$ J = w_1 \frac{|a_{limit} - a_{bef}|}{|a_{limit}|} + w_2 \frac{|a_{limit} - a_{aft}|}{|a_{limit}|} + w_3 \frac{L_p - L_{limit}}{L_{opt} - L_{limit}} \tag{10} $$

Where J is the evaluating indicator, a_{limit} is the maximum braking deceleration that is acceptable, a_{bef} is the deceleration selected before the lane change, a_{af} is the deceleration acquired in the acceleration field to ensure safety after the lane change, L_p is the longitudinal distance required for the lane change, L_{opt} is the temporary optimal longitudinal distance for the lane change given by trajectory planning method, L_{limit} is the shortest longitudinal distance needed but still within the limit of tolerance. w_1 and w_2 are weight coefficients responsible for measuring the importance of comfort in the braking process before and after a lane change, and w_3 is the weight coefficient measuring handling stability and comfort of the lane-changing process.

By calculating the evaluation value while giving priority to the span of a lane change, we can reach the best lane-changing scheme overall, which includes the braking deceleration and the time needed before the lane change, together with the longitudinal distance required by that lane change. When the optimal scheme is to change lane immediately, decision-making module sends the termination of that lane change to trajectory planning module and then make the change. When the better initial state for changing lane can be achieved by a period of braking, then prepare for the upcoming lane change by driving in the origin lane as instructed. The whole decision-making process can still be shown as in Figure 9.

Simulation with Polynomial Trajectory Planning

High order polynomials have properties of at least two order continuity. Taking advantage of this strength, polynomial trajectory planning methods obtain smooth trajectories while considering constraints of both start and end states. Instead of emphasizing the innovation of trajectory planning methods, this paper hopes that by choosing a stable and reliable planning method to cooperate with the proposed lane changing decision-making method, a complete decision-making & planning system can be formed, in which a vehicle can determine the timing of a lane change and obtain a feasible trajectory. Therefore, this paper only adopts the five order polynomial trajectory planning method to generate feasible trajectory clusters and extract the temporary optimum trajectory and the shortest trajectory within comfort limit. This information will be delivered to decision-making modules to make a further decision. Afterwards, the overall optimum lane-changing state will be fed back to the trajectory planning method, then the ideal trajectory can be produced directly. After being processed by trajectory following module, this lane-changing trajectory can be finally achieved with actuators.

To verify the decision-making method proposed, a typical scenario shown in Table 1 is simulated. Simulation results are expressed in Figures 11-13, and decision-making parameters shown in Figure 11 are explained in Table 2.

In Figure 11 and Figure 12, we can see that at the beginning of simulation, host vehicle is in the lane 1 (Lane = 1), acceleration calculated in the acceleration field approximately equals 0, and the room for a possible lane change is rather ample, so there's no intention to change lane (LC_intention = 0), and no necessity to judge the feasibility for a lane change (LC_feasibility = –1). Therefore, the decision-making module gives the instruction to maintain the original state (Decision = 0).

When t = 2.197 s, distance to preceding vehicle hits the lane-changing intention reference line. Since velocity of front vehicle in the same lane is lower than that of host vehicle, and velocity of anterolateral vehicle is also higher than that, it's reasonable to generate the intention of changing lane (LC_ intention = 1). When feasibility is checked and confirmed (LC_feasibility = 1), decision-making module gives the instruction to change lane (Decision = 3). From then on, the host vehicle is seen as in both original and target lane at the same time (Lane = 12).

TABLE 1 Simulation scenario

Parameter	Value
Initial velocity of the host vehicle	40 km/h
Initial velocity of the front vehicle in the same lane	35 km/h
Initial distance to the front vehicle in the same lane	10 m
Initial velocity of the front vehicle in adjacent lane	45 km/h
Initial distance to the front vehicle in adjacent lane	5 m
Initial velocity of the rear vehicle in adjacent lane	45 km/h
Initial distance to the rear vehicle in adjacent lane	20 m

FIGURE 11 Decision-making parameters.

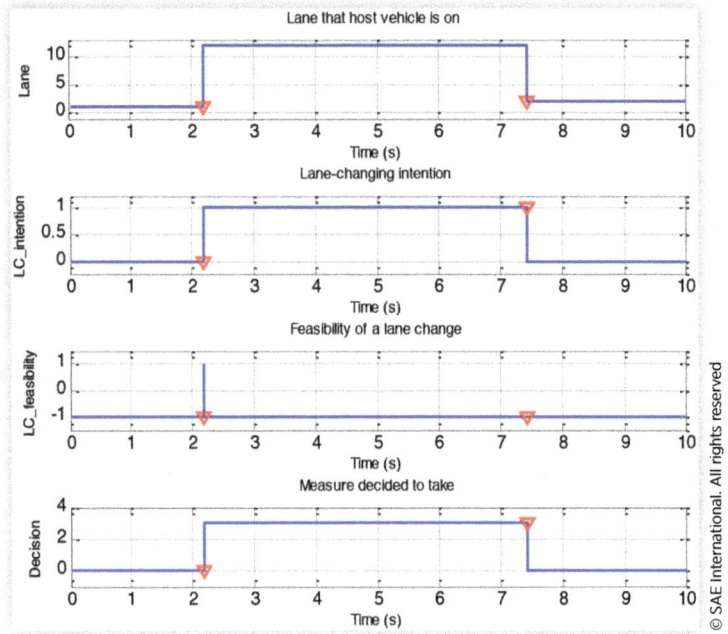

FIGURE 12 Relationships with the front vehicle in original lane (Blue) and adjacent lane (green).

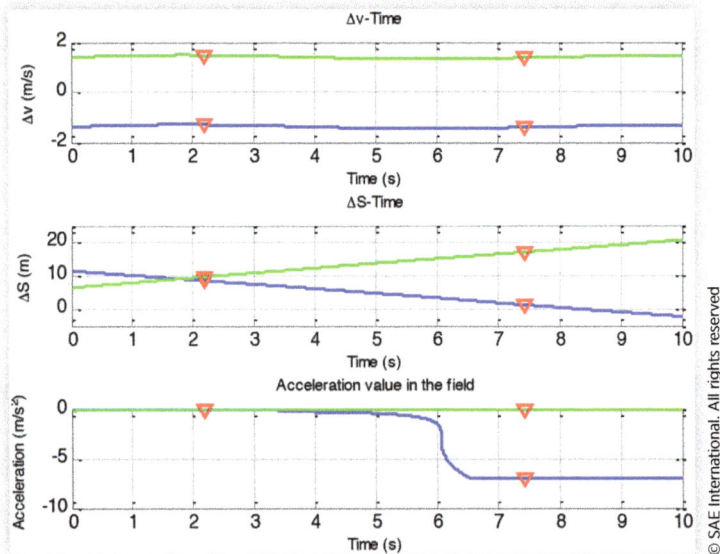

FIGURE 13 Some kinematic and kinetic parameter.

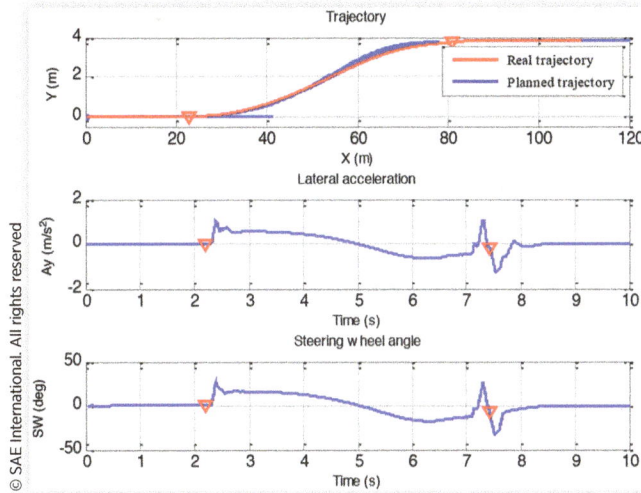

TABLE 2 Decision-making parameters in Simulation

Parameter	Meaning	Description
Lane	Lane that host vehicle is on	1: host vehicle is on the original lane
		2: host vehicle is on the adjacent lane
		12: host vehicle is changing lane
LC_intention	Lane-changing intention	0: Without intention to change lane
		1: With intention to change lane
LC_feasible	Feasibility of a lane change	-1: There's no need to judge feasibility
		0: Unfeasible to change lane
		1: Feasible to change lane
Decision	Measure decided to take	0: keep the state unchanged
		1: Brake
		2: Prepare for a lane change
		3: Change lane

The whole lane-change process lasts for 5.22 s, during which the host vehicle's trajectory, lateral acceleration, and steering wheel angle are shown in Figure 13. Figure 14 is the local enlarged drawing of the trajectory graph in Figure 13, it shows part of the periodic planning results, which exceed the lengths actually passed through by host vehicle in one planning cycle. As the results show, time for change is abundant, and the trajectory planned is quite smooth, which meets basic requirement of a lane change. Besides, in this case, the velocity of front vehicle is larger than that of host vehicle, so acceleration field gives the value of 0, which means there's no need to decelerate or accelerate. This is also in line with lane-changing rules.

On the whole, the decision-making method proposed in this paper achieves the goal of making a safe and comfortable lane change. As reference lines can be adjusted to adapt to different perceptual and driving habits, this method can be personalized.

FIGURE 14 Part of the periodic planning results.

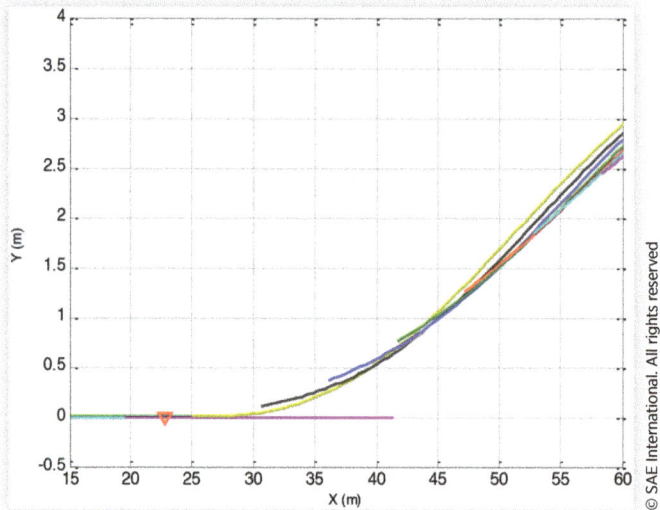

Summary

This paper proposed a new lane-changing decision-making method with consideration of safety, comfort, traffic efficiency and personality. First, we analyzed the process of braking and extended the emergency braking safety distance model to set up a braking acceleration field, then added the driving part to form a complete acceleration field that covers the common traffic environment. Second, we made a basic lane-changing decision-making model on the basis of acceleration field and some rules, which also caters for driver's habits, traffic efficiency and safety. Third, we extended the basic method and then velocity regulation was taken into consideration, which largely improve its flexibility. Finally, simulations with polynomial trajectory planning method were conducted to verify its superiority. As simulation results showed, the method presented in this paper can provide safety evaluation and give an optimum driving state on a structured straight road.

Contact Information

Dr. Bing Zhu
Professor
College of Automotive Engineering and State Key Laboratory of Automotive Simulation and Control
Jilin University, China
zhubing@jlu.edu.cn

Dr. Jian Zhao (Corresponding author)
Professor
College of Automotive Engineering and State Key Laboratory of Automotive Simulation and Control
Jilin University, China
zhaojian@jlu.edu.cn

Acknowledgments

This work is partially supported by National Key R&D Program of China (2016YFB0100904), National Natural Science Foundation of China (51475206, 51575225), and Jilin Province Science and Technology Development Plan Projects (20170101138JC).

References

1. Goodall, N., "Ethical Decision Making during Automated Vehicle Crashes," *Transportation Research Record Journal of the Transportation Research Board* 2424, no. 1 (2014): 58-65, doi:10.3141/2424-07.

2. Gipps, P.G., "A Model for the Structure of Lane-Changing Decisions," *Transportation Research Part B: Methodological* 20, no. 5 (1986): 403-414, doi:10.1016/0191-2615(86)90012-3.

3. Ahmed, K.I., *Modeling Drivers' Acceleration and Lane Changing Behavior* (Massachusetts: Massachusetts Institute of Technology, 1999).

4. Hidas, P., "Modelling Lane Changing and Merging in Microscopic Traffic Simulation," *Transportation Research Part C: Emerging Technologies* 10, no. 1 (2002): 351-371, doi:10.1016/S0968-090X(02)00026-8.

5. Seo, J. and Yi, K., "Robust Mode Predictive Control for Lane Change of Automated Driving Vehicles," SAE Technical Paper 2015-01-0317, 2015, doi:10.4271/2015-01-0317.

6. Chen, X., Miao, Y., Jin, M., and Zhang, Q., "Driving Decision-Making Analysis of Lane-Changing for Autonomous Vehicle under Complex Urban Environment," *Control and Decision Conference (CCDC)*, Chongqing, China, 2017, doi:10.1109/CCDC.2017.7978420.

7. Ding, J., Dang, R., Wang, J., and Li, K., "Driver Intention Recognition Method Based on Comprehensive Lane-Change Environment Assessment," *Intelligent Vehicles Symposium Proceedings, IEEE*, Dearborn, MI, 2014, doi:10.1109/IVS.2014.6856483.

8. Zhao, D., Lam, H., Peng, H. et al., "Accelerated Evaluation of Automated Vehicles Safety in Lane-Change Scenarios Based on Importance Sampling Techniques," *IEEE Trans Intell Transp Syst* 18, no. 3 (2016): 595-607, doi:10.1109/TITS.2016.2582208.

9. Zelek, J.S. and Levine, M.D., "Local-Global Concurrent Path Planning and Execution," *IEEE Transaction on System, Man and Cybernetics* 30, no. 6 (2000): 865-870, doi:10.1109/3468.895924.

10. Lavalle, S.M., "Rapidly-Exploring Random Trees: A New Tool for Path Planning," *Technical Report TR98-11, Department of Computer Science*, 12(4): 1-4, 1998.

11. Urmson, C. and Simmons, R., "Approaches for Heuristically Biasing RRT Growth," *2003 IEEE/RSJ International Conference on Intelligent Robots and Systems, 2003 (IROS 2003) Proceedings*, 2003, doi:10.1109/IROS.2003.1248805.

12. Ma, L., Xue, J., Kawabata, K., Zhu, J. et al., "Efficient Sampling-Based Motion Planning for On-Road Autonomous Driving," *IEEE Transactions on Intelligent Transportation Systems* 16, no. 4 (2015): 1961-1976, doi:10.1109/TITS.2015.2389215.

13. Jiang, Q., Deng, W., and Zhu, B., "Integrated Threat Assessment for Trajectory Planning of Intelligent Vehicles," SAE Technical Paper 2016-01-0153, 2016, doi:10.4271/2016-01-0153.

CHAPTER 8

14. Du, M., Chen, J, Zhao, P., Liang, H., Xin, Y., and Mei, T., "An Improved RRT-Based Motion Planner for Autonomous Vehicle in Cluttered Environments," *2014 IEEE International Conference on Robotics and Automation (ICRA)*, 2014, doi:10.1109/ICRA.2014.6907542.

15. Song, J.Z., Dai, B., Shan, E.Z. et al., "An Improved RRT Path Planning Algorithm," *Acta Electronica Sinica* 38, no. B02 (2010): 225-228.

16. Zhang, S., Deng, W., Zhao, Q., Sun, H., and Litkouhi, B., "Dynamic Trajectory Planning for Vehicle Autonomous Driving," *2013 IEEE International Conference on Systems, Man, and Cybernetics (SMC)*, 2013, doi:10.1109/SMC.2013.709.

Entropy in Reaction Space - Upgrade of Time-to-Collision Quantity

Vaclav Jirovsky
Czech Technical University in Prague

Today's vehicles are being more often equipped with systems, which are autonomously influencing the vehicle behavior. More systems of the kind and even fully autonomous vehicles in regular traffic are expected by OEMs in Europe around year 2025. Driving is highly multi-tasking activity and human errors emerge in situations, when he is unable to process and understand the essential amount of information. Future autonomous systems very often rely on some type of inter-vehicular communication. This shall provide the vehicle with higher amount of information, than driver uses in his decision making process. Therefore, currently used 1-D quantity TTC (time-to-collision) will become inadequate. Regardless the vehicle is driven by human or robot, it's always necessary to know, whether and which reaction is necessary to perform. Adaptable autonomous vehicle systems will need to analyze the driver's situation awareness level. Such knowledge can be enhanced by 2-D quantity, so called reaction space, and its entropy. The new approach defines a limit space, where ego vehicle or other vehicles can be present in the future specified by an amount of time. This enables the option of counting not only with braking time, but mitigation by changing direction is feasible. Opposed to TTC, considering time as an input is appreciated especially when switching from autonomous to manual driving. For such situation we observe two kinds of reaction spaces – one, connected with the requirements of autopilot, and second, resulting from the expected human reaction. Effects of entropy in 2-D reaction space are presented in the paper.

CITATION: Jirovsky, V., "Entropy in Reaction Space - Upgrade of Time-to-Collision Quantity," SAE Technical Paper 2017-01-0113, 2017, doi:10.4271/2017-01-0113.

Introduction

Current new vehicles are being more and more equipped with systems, which are autonomously intervening the driving process. EuroNCAP testing methodology even disqualifies vehicles without such systems from obtaining the all-stars rating from their test. The active and integrated safety systems together with advanced driver assistance systems (ADAS) could represent the first steps to creation of cooperative human-vehicle unit. Such unit should combine the best of the human with the best of algorithmic systems to form a symbiotic system achieving an outstanding performance in terms of safety. The main problem of current ADAS and similar systems originates in separate approaches to human and technology specifics – systems are designed to work without driver's intervention, while expecting the driver is being alert for the moment, when the system is incapable of processing correct reaction. The human nature does comply with such approach [1], thus we cannot await human's reaction in (un)expected situations. Systems themselves have to be designed in compliance with human's expectations. Systems, which are not designed to comply with human reactions, expectations and overall driving process, are already being implemented in large vehicle series. This leads to unwanted changes in behavior of the human drivers, who then become worse drivers. As an example of such system, we can take current adaptive cruise control, which definitely does not have the same approach to driving as a driver should have.

Driving in road traffic is highly contextual process. This means, that every driver shall adapt his behavior to the surrounding traffic and environment. The example of adaptive cruise control clearly demonstrates the incorrect behavior of the non-adaptive driver – current system only monitors the vehicle in front of the ego-vehicle and does not adapt to driver's will. This is proved i.e. during overtaking on highway when the road traffic is not locally fluent – when the driver ready for overtaking is not able to merge to the other lane in time expected by the adaptive cruise control, the ego-vehicle initiates the braking process. This induces new irregularity in otherwise predictable traffic, which human is forced to adapt to.

Current systems are monitoring only the time-to-collision (TTC) or time gap quantities, which are based on the nature of current sensor types. This quite unfortunate approach to the technology development is based on the limits of the current sensors and it is not reflecting correct driver's behavior and expectations. On the other hand, if we search the literature, we can see, that almost since the beginning of 1990s, it has been hard to find any scientific article about adaptive cruise control, which would not be implementing any type of v2x communication [2, 3]. It is obvious, that the ego-vehicle shall primarily adapt its behavior to the nearest vehicle, which is often the preceding vehicle. The problem in regular traffic arises, when the ego-vehicle's reaction (or either an action) is braking or any variant of deceleration, as this is the only reaction, which directly influences all traffic behind the ego-vehicle. Therefore, to keep the traffic as fluent as possible, we need to reduce the occurrences of deceleration. Reasonable reduction of braking events is performed via correct prediction of the behavior of surrounding traffic. This is very complicated for single sensor equipped vehicle and relatively easy achievable with functional v2x communication. In situations, where v2x communication is not available, continuous 360° environment monitoring around ego-vehicle and contextual data analysis and assessment can be sufficient in many situations.

With TTC and time gap approach, the ego-vehicle can expect only single reaction from the driver – braking. If the driver doesn't decelerate at specified moment, the vehicle classifies the situation as hazardous and proceeds with intervention (braking) by itself. When more sophisticated system is expected, like accident mitigation by steering or by combination of both, the system might expect the driver's reaction later (Figure 1). The problem originates in correct definition of situation-specific threshold – when the

FIGURE 1 Example of variant crash avoidance maneuver.

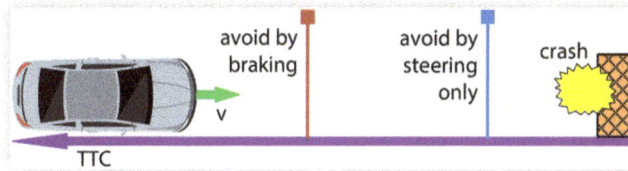

system can wait for the driver's reaction and when it has to engage the braking process. The proposed new approach of analysis of entropy in reaction space allows to estimate possible variants of behavior of ego-vehicle and also of other vehicles identified in the reaction space (or any monitored space). Therefore, the ego-vehicle does not have to brake, if the driver can choose an alternative reaction. The paper presents the approach to mathematical determination of entropy in 2-D road traffic environment. Entropy itself, as a scalar quantity, does not provide information about what kind of action or reaction is to be performed by ego-vehicle. It primarily denotes the need for reaction and specifies, how many types of reactions are feasible at any moment of driving. As the major input quantity is time, which limits the reaction space, the specific nature of the entropy allows to identify moments in future, when a reaction would be necessary. The specific trajectories and detailed reaction processes are not part of the work presented.

Definition of Reaction Space

Having sufficient reaction space is necessary to identify and analyze the critical situation, prepare an appropriate maneuver and perform it. Such space is primarily designated by the environment, in which the vehicle is currently moving. Let's define such reaction space ε as a quasi-2D area, whose theoretical boundary is limited by the position of the vehicle in the next n seconds. This space can be divided into two subspaces:

1. the first space in which the vehicle may be located during the maneuver to prevent collisions, i.e. either performs braking, an evasion maneuver or both - this space is bounded in Figure 2 by the red curves no. 1 and 5 and is denoted ε;

FIGURE 2 Schematic representation of reaction space (not in scale).

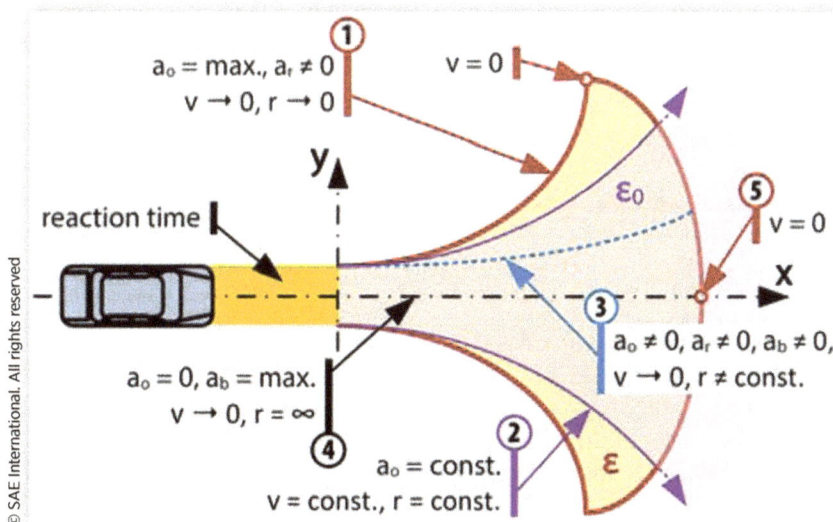

2. the second space in which the vehicle may be located while maintaining a constant driving speed - this space is shown in Figure 2 bounded by the purple line no. 2, and let it be called ε_0. Unlike for the space ε, the ε_0 is not restricted by the vehicle stopping distance and continues beyond the red line no. 5. During such maneuver the vehicle speed is not decreased by the vehicle resistance.

The diagram in Figure 2 indicates the character of the reaction space of the vehicle. Its boundaries are specified by several curves, which are determined by the dynamic capabilities of the vehicle. For the depicted reaction space, a simplified mass point model of the car is used. The extension to the model with dynamic properties of a particular car is not essential for the current work. The diagram indicates the main limit curve labeled 1-5, with the following specifications:

1. Curve no. 1, characterized by a maximum attainable combination of centrifugal acceleration and deceleration due to vehicle resistances a_r. It is therefore a curve whose radius of curvature r decreases with the length of the curve s and whose end point is defined by zero vehicle speed v. This curve is defined as clothoid (Euler spiral) [4] with the following derived parametrical definition:

$$
\begin{aligned}
x &= \int_0^T \cos\frac{\kappa}{a_r^3 t^2}\, dt \\
y &= \frac{b}{2} + \int_0^T \sin\frac{\kappa}{a_r^3 t^2}\, dt
\end{aligned}
\tag{1}
$$

where

- x is the direction of travel of the vehicle,

- a_r is the instantaneous deceleration of the vehicle,

- the variables t and T express either the time to stop or the time for which it is necessary to know the position of the vehicle in the future,

- and b is the width of the vehicle.

The vehicle's immediate deceleration a_r is given only by the vehicle resistances – air resistance, rolling resistance, cornering resistance or the resistance of slope. Constant κ is defined by the following equation:

$$
\kappa = \frac{1}{4} v_0 \mu_{max} g
\tag{2}
$$

where

- v_0 is the initial speed of the vehicle,

- μ_{max} is maximum achievable coefficient of adhesion between tire and road,

- g is the gravitational acceleration.

Further, the values of the length of the clothoid and of its radius are the input for the original equation of clothoid. Thus, by processing simple mathematical modifications we obtain the final equation Eq. (1) above. For the length s and radius r of clothoid, the following relationships are used:

$$s = \frac{v_0^2}{2a_r}$$

$$r = \frac{v^2}{a_c} = \frac{v^2}{\mu_{max}g} \tag{3}$$

$$dv = a_r dt$$

where

- s is the length of clothoid,
- r is the current radius of clothoid,
- a_c is the maximum centrifugal acceleration,
- v is the current speed of the vehicle.

2. Curve no. 2 is defined by the maximum centrifugal acceleration a_c too, but with the car driving at constant speed. Therefore, it encloses an area in which the vehicle is present when not responding to a critical situation. Such boundary curve is specified by constant radius, and therefore it is a circle:

$$x = \frac{v_0^2}{\mu_{max}g} cos\frac{\mu_{max}gt}{v_0}$$

$$y = \frac{b}{2} + \frac{v_0^2}{\mu_{max}g} sin\frac{\mu_{max}gt}{v_0} \tag{4}$$

3. Curve no. 3 is a trajectory of the vehicle when centrifugal acceleration reaches a maximum for a certain degree of braking force. In such case, it is also necessary to count with braking deceleration. It is again defined as a clothoid.

4. Curve no. 4, or rather a straight line, is the trajectory on which the maximum deceleration is achieved, while maintaining zero centrifugal acceleration. It is defined by a simple parametric equation:

$$x = \int_0^T \left(\mu_{max}g + a_r\right)t\,dt \tag{5}$$

$$y = 0$$

5. The reaction space is enclosed by the endpoints of all clothoids, covering all combinations of deceleration and cornering, in which the vehicle has zero velocity. The maximum value of the combined acceleration, or deceleration, is based on the so-called G-G diagram. It characterizes the contact between the tire and the road surface (see Figure 3) expressed in terms of achieved centrifugal acceleration ac, braking and resistive deceleration $a_b + a_r$ and forward acceleration a_a, which is primarily limited by a power of the vehicle. Such curve is clothoid again, this time symmetrical about the x-axis and passing through its zero point.

The resulting reaction space is bounded by the curves specified above. The calculation includes several known quantities (initial velocity, vehicle deceleration, physical constants and dimensions of the vehicle), then one

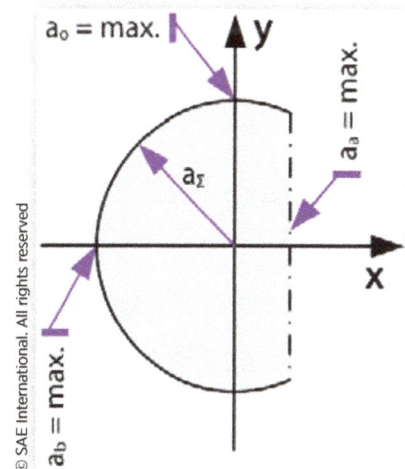

FIGURE 3 G-G diagram.

FIGURE 4   Example of intersection of two reaction spaces (not in scale).

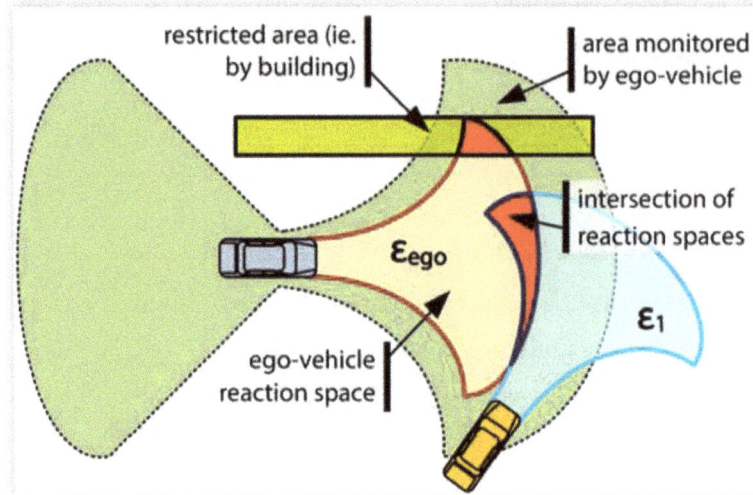

quantity depending on the instantaneous vehicle speed (the deceleration based on vehicle resistances) and finally a parameter identifying the required scope of knowledge about the movement of the vehicle in the future, which is time.

Relating the reaction space to the time base is particularly advantageous in terms of easy measurement. Calculating the probability of intersection of multiple reaction spaces of other vehicles and road users in the ego-vehicle's vicinity also allows more precise prediction of hazardous situations than the application of TTC. Furthermore, the reaction space can be simply extended to the monitored area by changing the time parameter and increasing the offset from the axis of the ego-vehicle. Sufficient size of the monitored area is one of the major variables influencing the ability to clearly predict future behavior of the ego-vehicle's environment. The size of monitored area is also tightly connected with the reaction time of the driver and vehicle systems [5].

The reaction space is closely related to speed before reaction. As the reaction space gets smaller, the limit speed for performing an accident avoidance maneuver also decreases, and vice versa. Nevertheless, it is necessary to know whether and how it is possible to perform a reaction in the reaction space and if the reaction is absolutely necessary. It is always advisable to consider whether mitigation of anticipated risks could cause more damage than the accident or risk itself. Such analysis should be essentially performed in terms of minimizing injuries, regardless of the material damage.

Generally, the reaction space itself is always influenced by the road environment and current traffic situation. In some cases, braking may be the only feasible reaction. Therefore, the reaction space can be used not only from the vehicle's perspective, but it can also be beneficial for the road design when determining i.e. the maximum allowable speed. When the application of dynamic traffic control takes control, the traffic density should be the main input parameter for the reaction space.

Interactions and Entropy

Interactions between all entities in the transportation system occur on various levels. We can distinguish two levels:

- communication – contactless interaction, low energy exchange;
- accident – contact interaction, high energy exchange.

Whether and what type of interaction will occur depends on the dynamic order of entities in the transportation system. This orderliness, resp. disorderliness, is determined by the probability of occurrence of each entity in the transportation system and the probability of the state of the entity. To quantify such orderliness, definition of entropy can be used. However, the current concept of entropy in transportation systems is primarily connected with urbanism [6, 7], particularly with:

- the human needs for transport within city area;
- the social stability and adaptation capability of the city to externalities;
- socio-economic modeling of transport use.

The author's application of entropy in prediction of possible interactions between entities in the transportation system is entirely new.

Definitions of entropy in various disciplines do not differ in principle – entropy is always a product of the probability of a specific state of system's entity occurrence and the quantity defining the physical nature of the system. Most common and universal entropy definition was published in 1870s by J. Willard Gibbs, utilizing the principles of statistical thermodynamics:

$$S = -k \sum_{i=1}^{n} P(x_i) \cdot \ln P(x_i) \tag{6}$$

where

- $P(x_i)$ is probabilistic function of system's entity occurrence in specific state x_i;
- n is the total number of possible states;
- k is a constant defining the physical nature of the system.

For the information theory, entropy was defined by Claude Shannon in 1949 as a value representing the maximum amount of information I in a message. The information transferred by message was defined by Ralph Hartley in 1927 by logarithm of inverted value of probability. Therefore, replacing k with $log_2 e = 1/ ln2$ specifying the bit quantity we obtain entropy in information system:

$$H = -\sum_{i=1}^{n} P(x_i) \cdot log_2 P(x_i) \tag{7}$$

As a definition of entropy close to information theory and reflecting the nature of transportation system, we can assume John von Neumann's entropy in quantum physics, introduced in 1932:

$$S = -Tr(\rho \cdot ln\rho)$$
$$\rho = |\psi(r, t)|^2 \tag{8}$$

where ρ is a density matrix describing the state of the system, which is a statistical data set of all quantum states defined by complex wave function ψ based on position r and time t. Function $Tr(A)$ is the trace of the matrix $A = \rho \cdot ln\rho$. The density matrix is a container of all system states with defined probability of their occurrence. Each of the system state is present in the matrix just once. Quantum states of the entities could analogically represent the states in transportation system.

For application of entropy in the interactions between the entities of transportation system, it is necessary to define the boundary conditions, where the analysis of entropy could be performed. In case we would like to determine the entropy of the road transportation as a whole, we would need to know all the entities in the system together with the probability of their mutual interactions. This is practically not feasible, therefore we have to define a closed area near ego-vehicle, where we are able to define the entropy. As such we use the above defined reaction space and the monitored enhancement of the reaction space. The reaction space refers to area, in which the correct reaction has to be always performed. Even though the reaction space does not have to be always constant, we can substitute the constant k in Eq. (6) with the area of reaction space.

Let Φ define a vector of presence of entity in the vehicle's close environment (reaction space), $P(\Phi_i)$ the probability of presence of the ith entity, ψ the vector of j unique interactions between i^{th} entity and ego-vehicle, and $P(\psi_j)$ the probability of every j^{th} interaction. Let's further define that the probability of interaction $P(\psi_j)$ is null in case, when the probability of presence of the entity $P(\Phi_i)$ is null. This defines the nature of conditional entropy. Therefore, we can write the entropy in vehicle's reaction space $\varepsilon \in (0; \infty)$:

$$s(\Phi, \Psi) = s(\Phi) + s(\Phi | \Psi)$$

$$s(\Phi, \Psi) = -\frac{1}{\varepsilon} \sum_{i=1}^{n} \sum_{j=1}^{m} \left(P(\Phi_i) ln P(\Phi_i) + P(\Phi_i) ln P(\Psi_j) \right)$$

(9)

The value of the entropy describes the complexity of interactions between entities in the reaction space. The higher the entropy, the more complex or earlier reaction of ego-vehicle is needed. At last, it is necessary to mention that instead of the reaction space ε, any other space can be used, thus entropy of such space could be determined.

The real life application then builds up on the nature of the entropy value. Large reaction space combined with low number of potential interactions leads to relatively large number of trivial solutions to pass through the defined space. On the other hand, minimizing the reaction space while maintaining the number of interactions, the number of solutions is also decreasing, but their complexity is rising. The side effect produced by the size of observed reaction space leads us to an idea about computing power needed to produce the vehicle's reaction. In addition, similar value of entropy for two different reaction spaces characterizes not only the similar need for the computing power, but also the similar complexity of ego-vehicle reaction.

The entropy of reaction space can gain values between 0 and infinite. The null value is reached at times, when the probability of presence of each entity together with probability of all the interactions is 1. This refers to situation, where there is only one entity in the reaction space and only one reaction is feasible.

Two other specific values of entropy can be further determined when:

1. probability of presence of each entity is $P(\Phi_i) = 1$ and the probability of each interaction $P(\psi_j)$ is not zero - then the solution simplifies to:

$$s(\Phi, \Psi) = -\frac{1}{\varepsilon} \sum_{j=1}^{m} ln P(\Psi_j)$$

(10)

2. presence of entity or interaction which has not been expected occurs, then the probability of such is zero, therefore the entropy is not defined for such situation.

For the second case, we can unfortunately find final reaction space so small, that his contribution to generation of reaction will become irrelevant.

It is unfortunate, that the dynamic nature of opened semi-self-organized road transportation system has a significant influence on the reaction space. However, the reaction space is never a constant and in some cases, it might be even undetermined. It further confirms the specifics of transportation system, which cannot be made adequately autonomous without centralized management and control and without knowledge of all situations, which might occur in the system.

Application Examples

The article describes two quantities – reaction space and entropy – which can be beneficially applied in the road transportation system. Combination of both provides new quality of knowledge about the state of the transportation system in specific area. Specific application of "entropy in reaction space" quantity predefines also the method for calculation, which, in some cases, can be simplified to prediction of presence of multiple entities in a specific differential area. We can see two main opportunities for application of the presented quantity:

- prediction of number and complexity of vehicle trajectories and prediction of optimal trajectory itself;
- prediction of accident probability in specific area.

The first application in prediction of ego-vehicle trajectory originates in similar process every driver does (or should do) – at a time before reaching a situation, driver classifies the probability of behavior of each entity entering the situation and chooses a trajectory according to this analysis. In terms of calculation of entropy in reaction space, the final trajectory is defined by connecting the optimal entropy levels (which does not have to be the minimal level) in each integration step. The process results at single state are schematically described in the Figure 5. As the entropy in each integration step changes in time, we can obviously incorporate time-based prediction for the determination of the final trajectory.

The diagram represents two vehicles and a pedestrian, each with their corresponding reaction space defined by the same time constant. This means that every entity can reach the border of its reaction space in the same amount of time. Colors in each of the boxes (integration steps) in the ego-vehicle reaction space represent the probability of its occupation by any of the entities at a corresponding time – red states for maximum, green for minimum. Unoccupied trajectory for the presented single state is then indicated

FIGURE 5 Entropy level in reaction space example (not in scale).

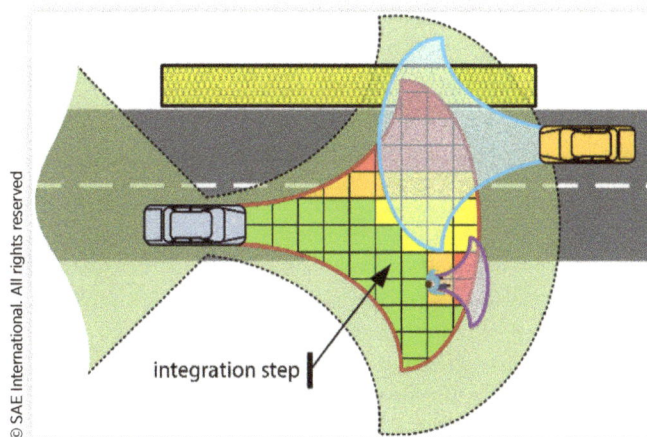

integration step

by the green area. We can clearly see the evidence for application of time-based prediction in order to keep the traffic flawless (moving yellow area). But, if the orange vehicle would unexpectedly change its trajectory in a negative manner, the ego-vehicle already knows the evasive safe route.

The second example illustrates the application on traffic management level. Based on the knowledge of presence and speed of each entity in centrally managed road traffic, locations with rising entropy can be identified. Thus, traffic optimization to maintain reasonable level of entropy can be performed.

The application in driver-controlled vehicle does not definitely imply the need for v2x communication. An ADAS or integrated safety system based on the knowledge of entropy in specified area around the vehicle can predict the driver's maneuver, thus create at least a basic concept of driver-vehicle symbiotic unit [5].

On the other hand, when looking on the conceptual application in road traffic management system, the need for functional v2x system or for a dense sensor network in road structure looks obvious. Current semi-self-organized road traffic is based on a sort of centralized management, which is not always obeyed (and it often cannot be, mainly due to the operational and system design errors). The traffic management will always be a combination of several communication technologies, which will provide the road users with necessary information how to behave in the traffic. The knowledge of entropy changes in the system can create a new approach to traffic management, especially in the mixed type of traffic, where autonomous and driver controlled vehicles will co-exist side by side.

Conclusion

Moving in the road traffic is highly contextual process. The road transportation is definitely one of the most interdisciplinary areas human was ever confronted with. On contrary to air or naval traffic, which, in terms of specification of entities and their interactions, are quite describable, the road transportation system is very opened system. While the human has created at least the majority of air or naval traffic, ground or road transportation system was and still is only being modified by human. This happens because it already existed long time before human. Many of the trails were created ad hoc by people or animals and nothing indicates that they will stop doing it in the future. It can be stated that people using the ground transportation system subconsciously realize probabilistic computations of presence of entities and of their interactions. But, they gained a significant advantage in situations, where the entropy is undefined. This advantage is in the ability of abstraction and genetically coded experience, which is very difficult to transfer to the "computer brain", as it is very fuzzy and the border between the information and noise is ambiguous.

The proposed reaction space and its entropy is trying to better reflect the correct driving behavior of a human driver – it analyzes behavior not only of the vehicle in front of the ego-vehicle, but also the behavior of other surrounding traffic depending on the geometrical configuration of the monitored space. While the reaction space strictly defines an area of all possible locations of ego-vehicle in future determined by specific amount of time and calculation of entropy in this space is critical, the monitored space is an area in which we can observe behavior of other entities and where the entropy predicts the need for reaction with certain probability. Both of these values can be beneficially applied in systems performing also other autonomous reactions than deceleration.

Contact Information

Václav Jirovský, Ph.D.
Vehicle safety researcher
Czech Technical University in Prague
Faculty of Transportation Sciences & Faculty of Mechanical Engineering
Tel.: +420 603 755 524
x1.jirovsky@fd.cvut.cz

References

1. International Harmonized Research Activities (IHRA), UNECE, "Design Principles for Advanced Driver Assistance Systems: Keeping Drivers In-the-Loop," International Harmonized Research Activities (IHRA), UNECE, 2010.

2. Desoer, C.A. and Sheikholeslam, S., "Longitudinal Control of a Platoon of Vehicles with No Communication of Lead Vehicle Information: A System Level Study," Path technical memorandum 91-2, 1991.

3. Desoer, C.A. and Sheikholeslam, S., "Longitudinal Control of a Platoon of Vehicles," *American Control Conference*, May 23-25, 1990.

4. Rektorys, K., *Přehled užité matematiky*, 2nd ed. (Prague: Nakladatelství technické literatury, 1968), 1140.

5. Jirovský, V. and Cappas, A.T., *Design Principles of Post-Autonomous Vehicles* (Brussels: EARPA FORM Forum, 2016).

6. Cabral, P., "Entropy in Urban Systems," *Entropy* 12 (2013): 5223–5236.

7. Wilson, A., "Entropy in Urban and Regional Modelling: Retrospect and Prospect," *Geographical Analysis* (2010): 364–394.

8. Camazine, S., Deneubourg, J.-L., Franks, N.R, and Sneyd, J., *Self-Organization in Biological Systems* (Oxfordshire: Princeton University Press, 2003), 560.

9. Jirovský, V., "Metodika hodnocení systémů integrované bezpečnosti osobních automobilů (Evaluation methodology of testing personal vehicle integrated safety systems)," Dissertation thesis, Czech Technical University in Prague, Prague, 2015.

10. Parasuraman, R. and Miller, C.A., "Who's in Charge?: Intermediate Levels of Control for Robots We Can Live with," *Proceedings of the IEEE International Conference On Systems, Man and Cybernetics*, Washington, DC, 2003.

11. von Bertalanffy, L., *General Systems Theory* (New York: George Braziller Inc., 1968).

A Maneuver-Based Threat Assessment Strategy for Collision Avoidance

Yaxin Li, Weiwen Deng, Bohua Sun, and Jian Zhao
Jilin University

Jinsong Wang
General Motors LLC

A dvanced driver assistance systems (ADAS) are being developed for more and more complicated application scenarios, which often require more predictive strategies with better understanding of driving environment. Taking traffic vehicles' maneuvers into account can greatly expand the beforehand time span for danger awareness. This paper presents a maneuver-based strategy to vehicle collision threat assessment. First, a maneuver-based trajectory prediction model (MTPM) is built, in which near-future trajectories of ego vehicle and traffic vehicles are estimated with the combination of vehicle's maneuvers and kinematic models that correspond to every maneuver. The most probable maneuvers of ego vehicle and each traffic vehicles are modeled and inferred via Hidden Markov Models with mixture of Gaussians outputs (GMHMM). Based on the inferred maneuvers, trajectory sets consisting of vehicles' position and motion states are predicted by kinematic models. Subsequently, time to collision (TTC) is calculated in a strategy of employing collision detection at every predicted trajectory instance. For this purpose, safe areas via bounding boxes are applied on every vehicle, and Separating Axis Theorem (SAT) is applied for collision prediction, so that TTC can be calculated efficiently and accurately. Finally, a threat level index based on reverse TTC is used to quantize the threat degree of every

CITATION: Li, Y., Deng, W., Sun, B., Zhao, J. et al., "A Maneuver-Based Threat Assessment Strategy for Collision Avoidance," SAE Technical Paper 2018-01-0598, 2018, doi:10.4271/2018-01-0598.

traffic vehicle potential collision to the ego vehicle. Experimental data collected in field test are used in the model training, and the overall strategy is validated under PanoSim. Simulation results show that MTPM can accurately identify maneuvers such that the effective prediction on trajectories can be generated. TTC and threat index can be calculated timely. The proposed threat assessment strategy can not only assist collision avoidance systems to foresee dangerous situations, but also eliminate false alarm to certain extent.

Introduction

Conceptions and basic architecture of vehicle intelligence has been further clarified during its decades of development. Worldwide accepted standards, such as SAE J3016, classify automated driving systems by levels of automation, making the application scope of automated driving technology clear [1]. Automated driving technologies, or vehicle intelligence, are expected to reduce the risk of accidents, providing the driver a safer vehicle with more comfort and better performance [2].

Collision Avoidance (CA) technology, just like the collision avoidance behavior of human drivers, is a technology that can assist drivers to reduce the risk of accidents and ensure the safety of passengers. CA can be applied in both longitude and lateral control driving scenarios, which can be the basis of several commercial active safety technologies such as Forward Collision Warning (FCW), Autonomous Emergency Braking (AEB), and Lane Change Assist(LCA). In the implementation of CA, there are three questions must be answered: whether the vehicle is in threat of collision, when and how the threat is going to happen, and what to do with vehicles and drivers to avoid or reduce collision? The third question, related to decision making process in automated driving systems, is always solved by giving the driver warnings timely and properly, or implement timely and effective control of the vehicle. The first two questions need understanding and modelling for driving environment basing on available environment state information. In this case, environment model is basically retrieved by understanding the situation, namely situation awareness, and the whole process is called situation assessment. The definition of situation awareness is:" The perception of elements in the environment within a volume of time and space, the comprehension of their meaning, and the projection of their status in the near future" [3]. Furthermore, since CA focuses on the occurrence or not of the collision, research on collision threat assessment based on situation assessment is necessary.

Performing collision avoidance behavior in time and reducing false alarm rate are always what CA technologies care. Research shows that 60% of collisions can be avoided with a warning 0.5 s in advance, and the percentage becomes 90% for 1 s beforehand [4]. Long-term trajectory prediction can help expanding beforehand time span, and computational efficiency is also crucial for the time span. For situation awareness, there are three major sources of uncertainty, which are state estimates, driver maneuvers and driving maneuver executions for the realization of the maneuver [5]. Maneuver-based threat assessment can not only eliminate the uncertainty brought by driver's maneuver, but also making trajectory prediction reliable to strive for longer prediction time span.

In simple application scenarios such as rear-end collision, a widely used way to assess collision threat is employing safety distance model [6] or time-to-collision (TTC) model [7,8] to define whether a situation is dangerous or not. This method owns relatively low time complexity, but cannot handle arbitrary collision. In state-of-art threat assessment methods, the trajectories of vehicles are always predicted to adapt more scenarios. The task for trajectory prediction is predicting future motion state of vehicles. Motion prediction method can be clustered into three categories: physical-based, maneuver-based and interaction-aware [9]. Physical-based models including dynamic model and kinematic model which are applied in [10], and models with noise considered, implemented in Kalman Filter or Unscented Kalman Filter, as [11] has shown. The research in [9] emphasizes that although physical-based models can achieve the highest computational speed, they are only limited to short-term predictions. Interaction-aware models, opposite to physical-based method, may bring the highest time complexity. Maneuver-based method can bring the best trade-off between long-term prediction precision and computational efficiency.

The maneuver-based threat assessment method in [12] define maneuvers with Kamm's circle, which reflects the adhesion ability of the vehicle's tire, so that a branch of possible trajectories can be predicted with kinematic model. A collision index is defined by vehicle states and pre-defined shapes, and is described in a collision index matrix by every prediction time instant. The method serves CA decision making more than threat prediction. The research in [13] assumes the driver may maintain current behavior in the near future, consequently applies a lane-keep driving model with a state predictor to generate predicted trajectory. However, CA systems should also be effective in abnormal driving situation, and the accuracy of lane-keep model to describe other maneuvers such as lane changing is not enough. It is evident that maneuvers should be clearly recognized. During threat assessment, maneuvers can be recognized in several ways, by comparing the parabolic model of the lane and the possible trajectories [14], or by inferring ways like Bayesian Network [15]. However, the methods mentioned assume vehicles' shape as points, losing sight of the fact that orientation and profile of objects can also impact assessment accuracy.

This paper focuses on an effective method that can quantitatively assess collision threat timely and credibly. To achieve this goal, both maneuvers and vehicles' shapes are considered. The presentation of our maneuver-based threat assessment strategy will start from the overall structure, giving a review of the way in which threat assessment is implemented. Then maneuver-based trajectory prediction model (MTPM) will be shown, followed by collision prediction method. The last step of threat assessment, quantifying the threat, will also be described. An illustration of validation and simulation result is placed at the end of the paper.

Strategy Structure

Before introducing every stage of the maneuver-based situation assessment strategy proposed, the overall structure is shown in Figure 1. Both environment states and ego vehicle states are required to assessment the collision threat in the driving environment. Among all the possible traffic participants, traffic vehicles' states are considered, while the states of pedestrian and static obstacles are not included in this paper.

Basing on the states of traffic and ego vehicle, the future trajectory of those vehicles can be separately predicted in maneuver-based trajectory prediction model (MTPM). In MTPM, long-term future trajectories are predicted with two procedures, maneuver recognition and corresponding kinematic trajectory generation. Trajectories are described as time series, including future vehicles' states in prediction time horizon. To finish maneuver recognition work, Hidden Markov Models with mixture of Gaussians

FIGURE 1 Structure for maneuver-based threat assessment.

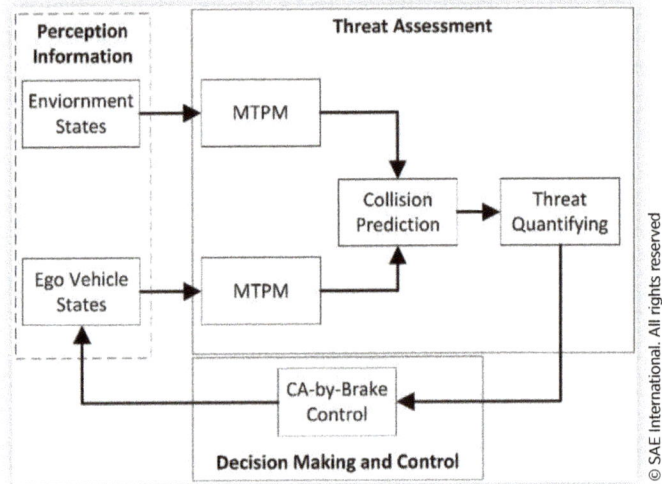

outputs (GMHMM) is employed. Collision prediction algorithm is applied to every time instant of future trajectories, so that time-to-collision (TTC) can be calculated with high accuracy. In near area collision accident, the shape of vehicles cannot be ignored, so that vehicles are presented in bounding boxes. Separating Axis Theorem (SAT) in computer graphics assures that collision prediction algorithm is done in an accelerated way. According to inverse TTC, threat is quantified to be the basis of decision making and control. In this paper, we care about CA by brake most, hence steering assistance is not discussed. In the Simulation and Test Result part, experiments with vehicle control will be present to prove the effectiveness if the strategy.

Maneuver-Based Trajectory Prediction Model

Maneuver Recognition with GMHMM

Hidden Markov Model (HMM) is a temporal probabilistic model, which can reveal how unobservable state variables and observable evidence variables of a system may change by time [16]. HMM has been widely used in speech recognition, and research shows that HMM also has great advantage in maneuver recognition. Compared to Logical Approaches, Markov Logic Networks, Bayesian Networks and Dynamic Bayesian Networks, HMM is capable of involving multiple objects and spatial-temporal dependencies situations [17]. HMM has tolerance to uncertainty, while reaching very high inference accuracy.

Two parameters should be defined at first:

N The number of hidden states, hence the state set of the model can be defined as $S = \{S_1, S_2, ..., SN\}$, with q_t denoting the state at time t.

M: The number of distinct observation symbols per state. Every symbol can be donated as $V = \{v_1, v_2, ..., v_m\}$.

An HMM can be put into a parameter set of three:

$$\lambda = (A, B, \pi) \tag{1}$$

where each parameter can be defined as follow:

A: The state transition probability distribution A = {a_{ij}}, where

$$a_{ij} = P\left[q_{t+1} = S_j | q_t = S_i \right], 1 \le i, j \le N \tag{2}$$

B: The observation symbol probability distribution in state j, B = {$b_j(k)$}, where

$$b_j(k) = P\left[v_k \text{ at } t | q_t = S_j \right], 1 \le j \le N, 1 \le k \le M \tag{3}$$

π: The initial state distribution π = {πi}, where

$$\pi_i = P\left[q_1 = S_i \right], 1 \le i \le N \tag{4}$$

Observation sequence can be named as O=$O_1 O_2 \ldots O_T$ In maneuver recognition application, observations should be continuous signal. In other word, M may be infinite. Therefore, Gaussian Mixture model is used to describe observation symbol probability distribution as a finite mixture model

$$b_j(O) = \sum_{m=1}^{M} c_{jm} N\left[O, \mu_{jm}, U_{jm} \right], 1 \le j \le N \tag{5}$$

where M becomes the number of mixture used in the model, c_{jm} is defined as the mixture coefficient for the mth mixture in state j, μ_{jm} is the mean vector and U_{jm} is covariance matrix for the mth mixture component in state j [18].

Hence a GMHMM can be configured as

$$\lambda = (A, c, \mu, U, \pi) \tag{6}$$

In MTPM, we care about lane keeping and lane changing maneuvers the most. Baum-Welch method is used to determine parameters of GMHMM, producing parameters for lane keeping and lane changing, λ_{LK} and λ_{LC}. After the model being trained, observations from sensors can be recognized as most possible maneuver in the forward-backward procedure. The maneuver with maximum log-likelihood will be chosen as the most possible maneuver.

The training datasets are attained from field test. OxTS RT3002 precision Inertial and GPS Navigation systems is used as sensor. With RT3002, both longitude and vertical kinematic states be measured, including position, velocity and acceleration. Figure 2 shows the process in which maneuvers re recognized. The parts in dash rectangle is finished offline based on driving datasets. After pre-processing, experiment data is divided into training datasets and testing datasets, taking up 75% and 25% of the total data separately.

Considering the detection function of popular sensors such as lidar and GPS, define the observation vector as

$$o_t = (y_R, v_y, a_y) \tag{7}$$

where y_R, v_y and ay are the lateral offset, velocity and acceleration of vehicle in road-fixed coordination. Before training and testing, the sensor data should be pre-processed. Besides filtered, data need to be windowed and scaled to reach higher model precision.

The reason for applying time window to data is that each maneuver owns its own duration time. Either the window being too long or too short will affect distinction of

FIGURE 2 Flow chart of maneuver recognition in MTPM.

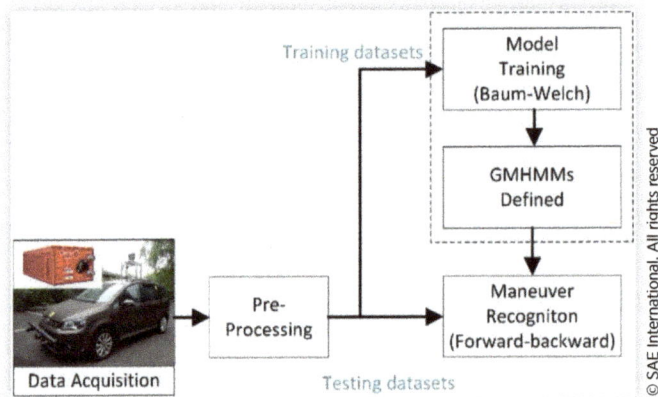

different maneuvers. Research shows 3 seconds can help the algorithm reach higher accuracy [19], and windows refresh every 100 ms. The data is labeled based on time windows. During data acquisition, drivers would carry out lane change maneuver after a start signal. Therefore, the starting point of lane change maneuver is defined as the moment the lateral offset to the destination lane is getting smaller, while the ending point is defined as the moment the lateral offset to the original lane stop getting larger. The first time window that contains the starting point of lane change will be labelled as lane change maneuver, until the last time window that exclude the ending point of lane change. The rule for time window and labelling will be illustrated in Figure 3. Red windows are labelled as lane change maneuver, while blue windows are labelled as lane keeping maneuver.

Data scaling can project sensor data into reasonable range, highlight data patterns in different maneuvers. Figure 4 shows the effect of data scaling. Choose the maximum value of every variable in observation vector to scale the vector

$$o_{t_{scaled}} = \left(y_{R\_scaled}, v_{y\_scaled}, a_{y\_scaled} \right)$$
$$= \left(\frac{y_R}{\max|y_R|}, \frac{v_y}{\max|v_y|}, \frac{a_y}{\max|a_y|} \right) \quad (8)$$

During the experiments, five drivers were asked to execute the same scenario for four times. At first, drivers would drive in lane. After receiving a signal for lane changing, drivers would implement a left lane changing maneuver, followed by lane keeping. Then in the same method, drivers would implement a right lane changing and keep in lane.

FIGURE 3 Data window labelling illustration.

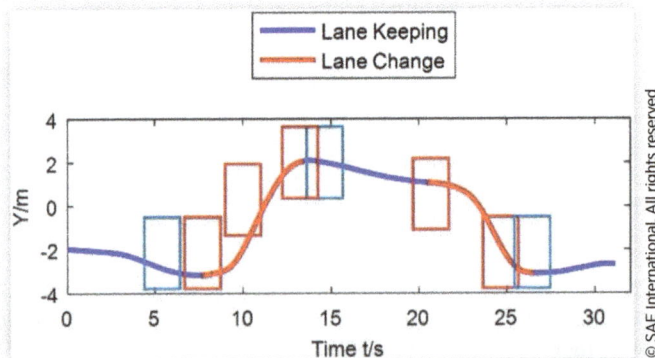

FIGURE 4 The influence of data scaling.

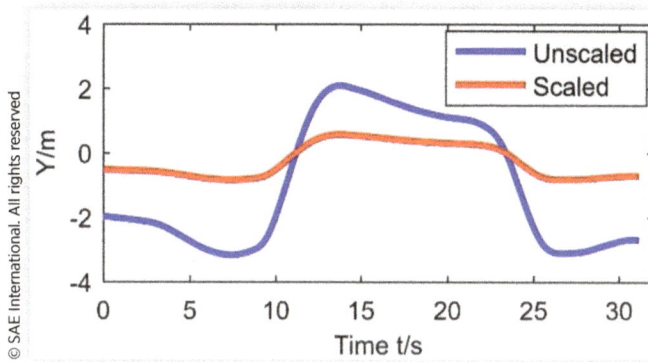

One scenario would last about 28 seconds, producing more than 250 valid labelled time windows each. In total, the whole experiment produced 12135 labelled time windows as samples, among which 7512 windows are labelled as "Lane Keeping" and 4623 windows for "Lane Change". GMHMMs are trained separately using their own maneuver samples. Defined by the number of windows whose maneuver type recognized coincides with their labels, the final recognition accuracy can reach 91.3%.

Taking the data in Figure 3 and Figure 4 for example. Comparing the labelled maneuver with recognized maneuver in Figure 5. Figure 5 shows that the shifting

FIGURE 5 Example for maneuver recognition results.

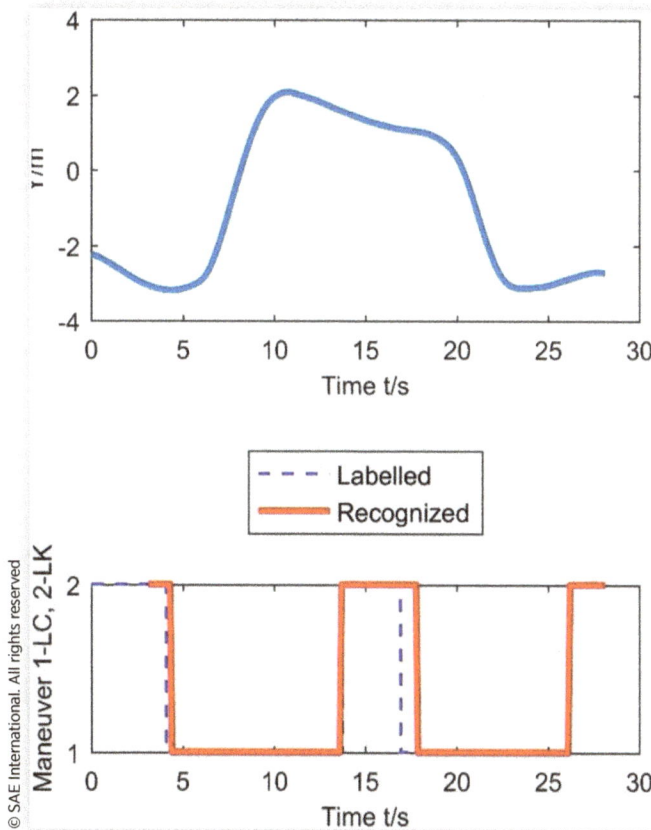

between maneuvers is always source of recognition errors. In this example, there are 251 windows in total, and 236 windows of them is recognized correctly, with an accuracy rate of 94.0%.

Trajectory Prediction

Based on the most possible maneuver recognized by GMHMMs, the trajectories of vehicles are predicted by corresponding kinematic model. Different kinematic models are applied on different maneuvers.

For lane keeping maneuver, Constant Acceleration model (CA) is used to predict trajectories, given by

$$
\begin{pmatrix} x_{t+1} \\ v_{t+1} \\ a_{t+1} \end{pmatrix} = \begin{pmatrix} 1 & T & \frac{1}{2}T^2 \\ 0 & 1 & T \\ 0 & 0 & 1 \end{pmatrix} \begin{pmatrix} x_t \\ v_t \\ a_t \end{pmatrix} + \begin{pmatrix} \frac{1}{2}T^2 \\ T \\ 1 \end{pmatrix} w_a \tag{9}
$$

where T is defined as sampling time, and w_a is process noise scaler. The original involving of noise scaler in CA model is to describe the overall processing and observing noise, which can help confirming the noise parameter in following trackers or describing state distributions. This paper is built in a deterministic way with the assumption that the signals involved have been filtered, so that the noise scaler has been set to zero. In this paper, driving in turns is not considered, and the velocity and acceleration in the model are both in longitudinal and lateral direction.

For lane changing maneuver, because of the method of labelling lane changing maneuver in maneuver recognition model, a combination of sine half-cycle Lane Change model (LC) [5] and Constant Velocity model (CV) is applied.

In LC model, the width of lane w_L should be predefined. Here define $w_L = 4$ m. With current path angle ψ_k and lateral lane offset y_k at moment k acquired from sensor data, the remaining maneuver length $l_{r,k}$ and longitudinal offset from maneuver starting Δx_k can be derived in equation (10)-(11). States in the next moment, y_{k+1} and Δx_{k+1} is calculated in equation (12)-(13).

$$
l_{r,k} = \frac{w_L \pi}{2 \tan \psi_k} \cos\left(\sin^{-1}\left(\frac{2 y_k}{w_L} - 2\right)\right) \tag{10}
$$

$$
\Delta x_k = \left(\frac{1}{2} + \frac{1}{\pi} \sin^{-1}\left(\frac{2 y_k}{w_L} - 2\right)\right) l_{r,k} \tag{11}
$$

$$
y(\Delta x_{k+1}) = \frac{w_L}{2} \sin\left(\frac{\pi}{l_{r,k}} \Delta x_k - \frac{\pi}{2}\right) + w_L \tag{12}
$$

$$
\psi_{k+1} = \tan^{-1} \frac{dy}{d\Delta x} = \tan^{-1}\left(\frac{w_L \pi}{2 l_{r,k}} \cos\left(\frac{\pi}{l_{r,k}} \Delta x_k - \frac{\pi}{2}\right)\right) \tag{13}
$$

FIGURE 6 CV is following LC to complete the total prediction.

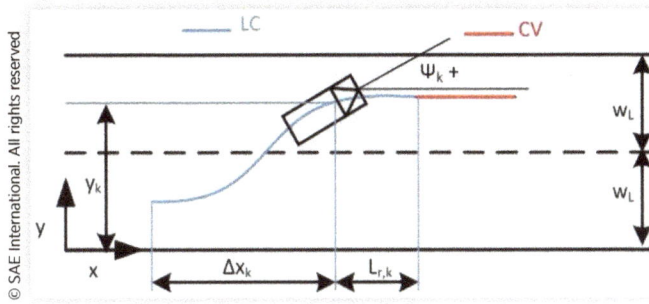

The longitudinal states is predicted by CV

$$\begin{pmatrix} x_{t+1} \\ v_{t+1} \end{pmatrix} = \begin{pmatrix} 1 & T \\ 0 & 1 \end{pmatrix}\begin{pmatrix} x_t \\ v_t \end{pmatrix} + \begin{pmatrix} \frac{1}{2}T^2 \\ T \end{pmatrix} w_v \qquad (14)$$

where just as CA, T is defined as sampling time, and w_v is process noise scaler.

The trajectory predicted by LC has a specific starting and ending. If the remaining trajectory is shorter than prediction time span, the blank prediction time will be filled by CV. The coordination definition and the combination of models is illustrated in Figure 6.

To show the validation of MTPM, we choose two segments on driving data, comparing the difference between predicted trajectory and the actual driving trajectory. Here we define the prediction time span is set as 3 seconds. The width of lane $w_L = 4$ m, sampling time T = 0.05 s, thus 60 steps of prediction are produced.

Figure 7 shows the segments chosen separately for trajectory prediction. The red segment is for prediction for lane changing maneuver, and the green segment if for prediction for lane keeping maneuver. In Figure 8, every red cross represents a prediction time step. CA in applied on both longitudinal and lateral motion for 3 seconds, realizing a good prediction result. In Figure 9 besides the driving data in blue line and the predicted trajectory in red cross by LC, a trajectory predicted by CA is presented in green circle. It shows that maneuver-based prediction can take advantage of motion characters of different in trajectory prediction, and can also extend the universality of trajectory prediction model.

FIGURE 7 The example maneuver and the two segments picked for validation.

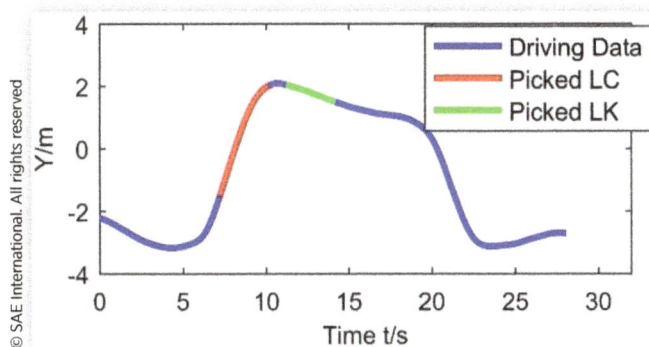

Comparison of predicted trajectory and actual trajectory for lane keeping maneuver.

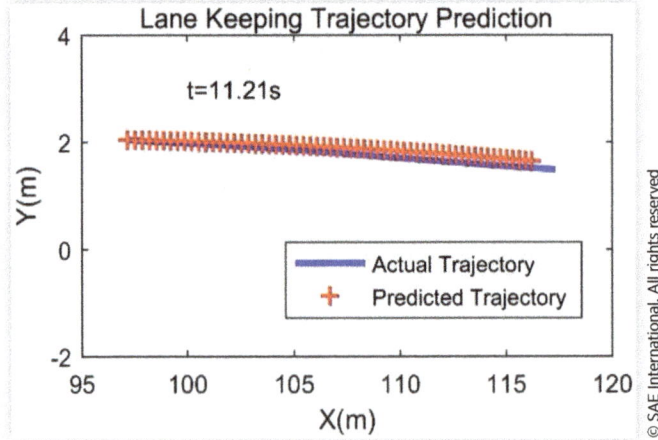

FIGURE 9 Example for maneuver recognition results.

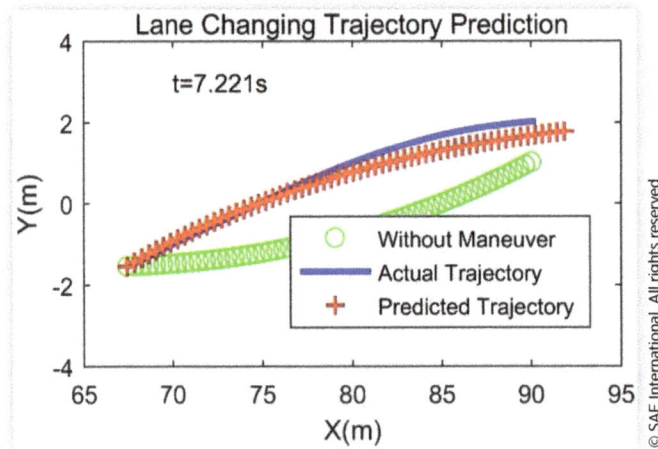

Collision Prediction Model

Given the trajectories generated by MTPM, a judgment and prediction of whether and when collisions would happen in the prediction time span need to be done. By applying collision judgment on every prediction time step, the collision time can be calculated. During the judgment, the shape and orientation of vehicles should be taken into account, while computational efficiency is also important. Vehicles are modelled in geometric method, and the collision judgment is finished in intersection test mothed in computer graphics.

Vehicles are abstracted into cuboid to describe their shapes. The collision is supposed to take place in two-dimensional space, so vehicles are modelled as Oriented Bounding Boxes (OBBs) in surface. An OBB, A, is described by its center point $\mathbf{a}^c$, and side directional normalized vectors $\mathbf{a}^u$ and $\mathbf{a}^v$. The distance between the center and the two sides are named as respective positive half-lengths, h_u^A and h_v^A.

To implement intersection test between two OBBs, using Separating Axis Theorem (SAT) [20] is an efficiency method. SAT represents that separating axes exists only if two convex polyhedrons do not intersect with each other. For two OBBs, only the sides directional axes need to be tested. If any of the four axes is separating axis, the two OBBs are not intersecting.

For two OBBs A and B, representing the ego vehicle and traffic vehicles separately, a vector connecting their centers is donated as **t**. The axis under testing is donated as **l**. The projection of the sum of half-length vector on **l** is donated as d_A and d_B. The definitions are illustrated in Figure 10.

Separating axis is tested by

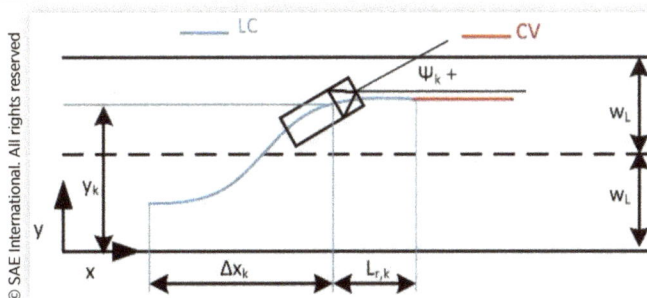

The illustration for OBB intersection testing.

$$\left| t \cdot l \right| > d_A + d_B, l \in \left\{ a_u, a_v, b_u, b_v \right\}$$

$$d_A = \sum_{i \in \{u,v\}} h_i^A \left| a^i \cdot l \right|$$

$$d_B = \sum_{i \in \{u,v\}} h_i^B \left| b^i \cdot l \right|$$

(15)

Intersection testing should be implemented at every prediction step by checking at most four sides. The first step that intersection happens will be treated as the moment collision takes place, with is also called TTC. To make the calculation faster, a simple judgment can be done before finding separating axis. If the distance between two OBBs is large and exceeds a threshold l_{far}, the intersection test should be done in a simpler way in

$$\left| t \right| > l_{far}$$

(15)

A test scenario is built as shown in Figure 11. Vehicles built in rectangles, whose length is 4 m and width is 2 m. Based on the scenario in Figure 9, a traffic vehicle is driving in the destination lane of ego vehicle. The traffic vehicle is driving in a velocity of 10 m/s, and decelerating in -2 m/s². Figure 12 shows the predicted trajectories of both vehicles. The trajectory intersects at the place x coordination is 90 m, but considering the shape of vehicles, collision take places earlier than that. Figure 13 tells that collision prediction algorithm indicate the collision will happen in 0.5 s, and the front of ego car will bump into the right side of traffic vehicle.

Test scenario for detection prediction.

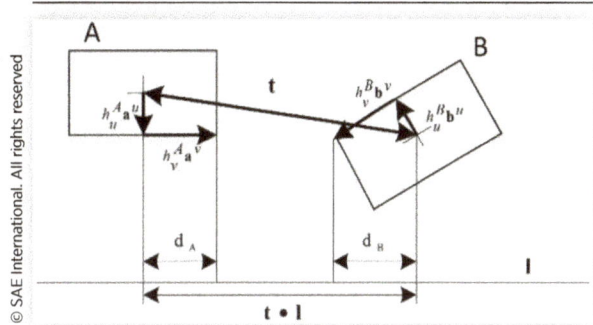

Trajectories of ego and traffic vehicle.

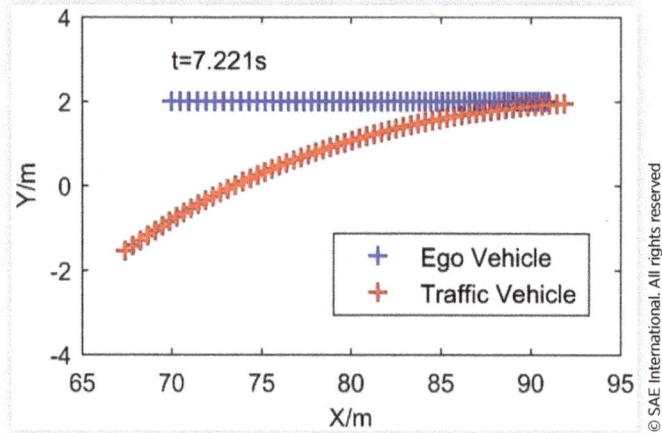

FIGURE 13 The moment that collision happens.

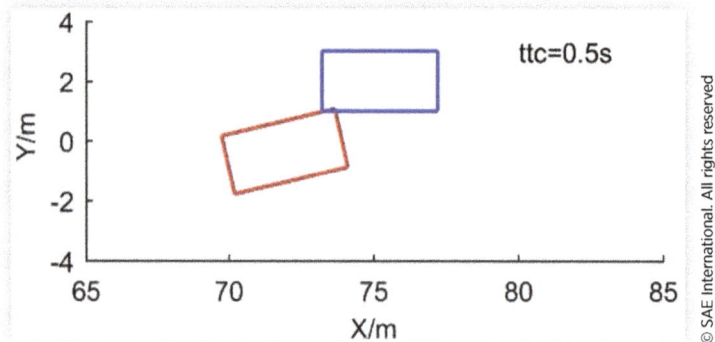

Threat Quantifying

The degree of collision threat is described in five levels: "Safety", "Warning", "Emergency warning", "Slight Interference" and "Harsh Interference". In the system, we care about CA by brake most. Since the driver behavior in emergency braking is related to reverse TTC, TTC^{-1}, an experimental hierarchical model is employed [21].

Assuming Level 1 represents 'Safety', the boundaries among levels can be presented as

$$TTC^{-1}_{boundary45} = \max\left(1.7609 - 0.0128v_f, 1.2\right)$$

$$TTC^{-1}_{boundary34} = \max\left(1.1184 - 0.0131v_f, 0.75\right)$$

$$TTC^{-1}_{boundary23} = \frac{1}{\max\left(1.1184 - 0.0131v_f, 0.75\right) + 1} \tag{16}$$

$$TTC^{-1}_{boundary12} = \frac{1}{\max\left(0.476 - 0.0134v_f, 0.2\right) + 1.22}$$

where v_f is the longitudinal velocity of vehicle in km/h. The levels are shown in Figure 14.

FIGURE 14 Illustration of hierarchical threat qualifying model.

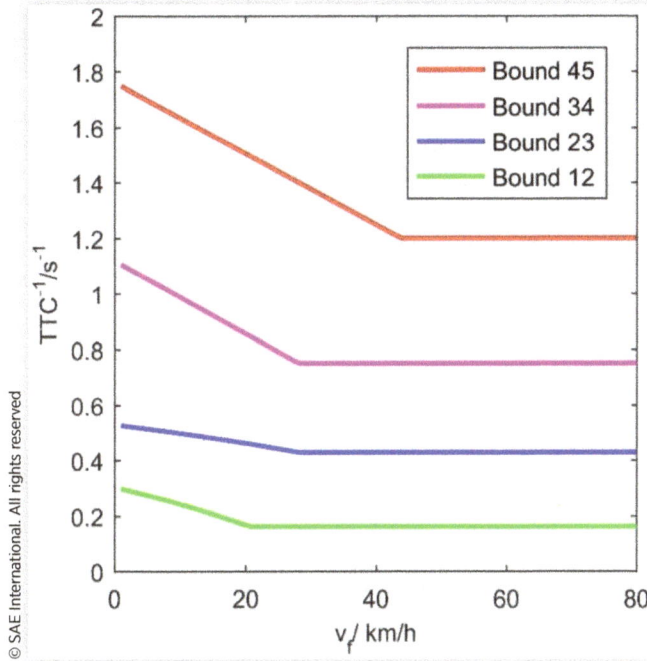

Simulation Results

The validation of the whole system need extreme scenarios like vehicle collision, thus the tests are finished in virtual driving enviornment. To test the maneuver-based threat assessment strategy, a simulation is implemented in MATLAB/ Simulink and coupled with the ADAS simulation environment PanoSim. In PanoSim, vehicles are modelled in vehicle dynamics, which have high simulation precision.

As it shows in Figure 15, the test field is designed as a double-lane straight road. Ego vehicle (red) is in the left lane, and traffic vehicle (gray) is in the right lane. Ego vehicle is keeping in lane with a constant speed of 12 km/h. Traffic vehicle is keeping lane at first, but it starts acceleration and changing into the left lane, producing a threat of collision. After lane changing, the velocity of traffic vehicle falls below 12 km/h and brings collision threat. The motion profiles of both vehicles are shown in Figures 17-19. The output of the strategy is threat level, so that interferences are not contained in the system.

The threat level is shown in Figure 20, while detected maneuver is in Figure 21, and TTC is shown in Figure 22. Assume that when TTC = 0 a collision would happen. In case of data overflow, TTC has an upper limit of 20s. In order to prove that TTC has been accurately calculated, a "ground-truth" collision symbol has been calculated by applying intersection test on the driving trajectories. The comparison of "ground-truth" collision symbol and calculated collision symbol is shown in Figure 23. Without interferences, the two vehicles are supposed to run into each other twice. The collision situations are shown in Figure 24. To validate the predict ability of the strategy proposed, the prediction time in advance is analyzed in Figure 25.

The error of this maneuver-based threat assessment strategy should be discussed. In Figure 25, the blue line represents the time that collisions are predicted to happen. In other words, it tells about what time the system "think" collisions would take place. The green line represents the ground-truth prediction about collisions, namely, the ideal

FIGURE 15   PanoSim panel and MATLAB/Simulink model.

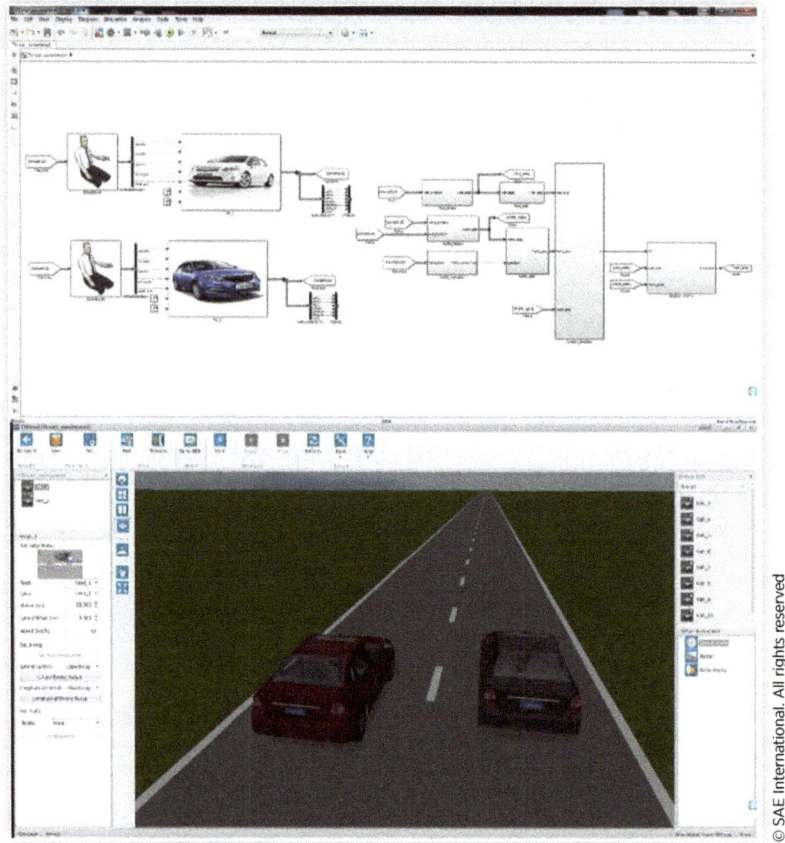

FIGURE 16   Illustration of simulation scenario.

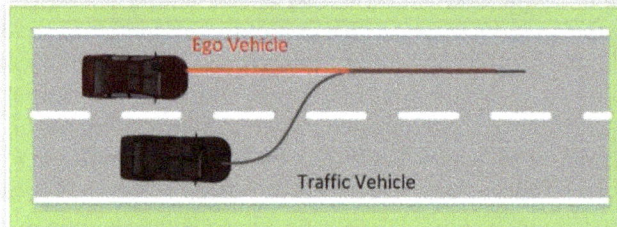

collision prediction results of CA. The black reference line shows up when the prediction timeline and ground-truth timeline are synchronized. Following the ground-truth collision symbol line in red, with the assist of the black reference line, we can find two parts that overlap with the black line. These two parts represents the period that collision is actually happening. The flat parts on blue and green line represents the beforehand time that a system can predict a coming collision. From Figure 25 we can tell that the system performs a little bit differently during predicting the twice collisions. For the first collision, the system predicted about 2 seconds in advance, and the collision time estimation results are close to the ideal results. However, the prediction for the second collision is about 3 seconds in advance, but the predicted collision time has larger error. The more near to collision time, smaller and more stable will the error be.

FIGURE 17 Longitudinal velocities of two vehicles.

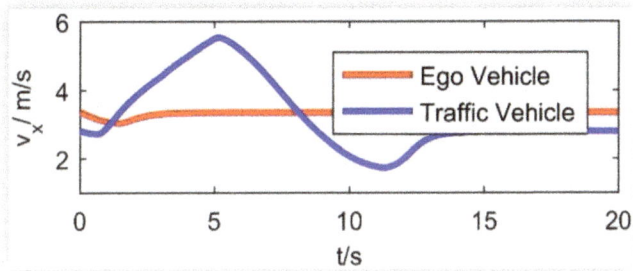

FIGURE 18 Y coordinates in global road-fixed coordinate of two vehicles.

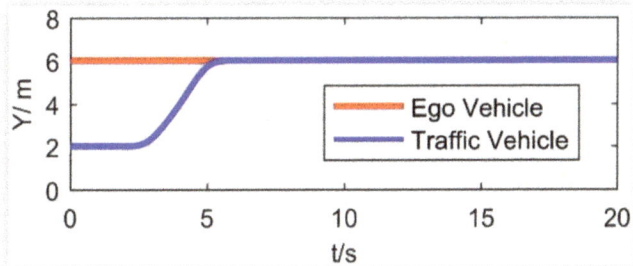

FIGURE 19 The trajectories of vehicles in global road-fixed coordinate.

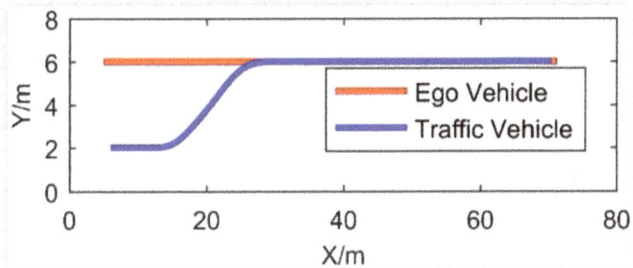

FIGURE 20 Profile of treat level.

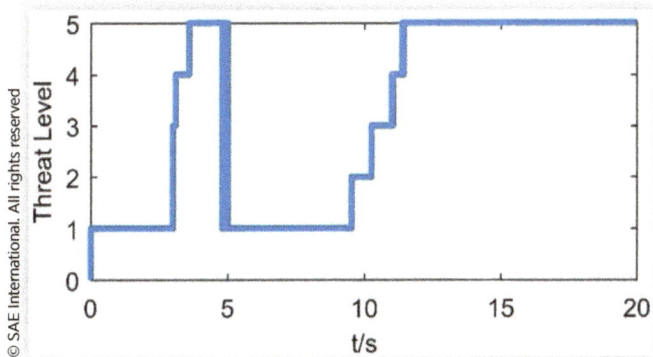

FIGURE 21 Maneuver recognition results.

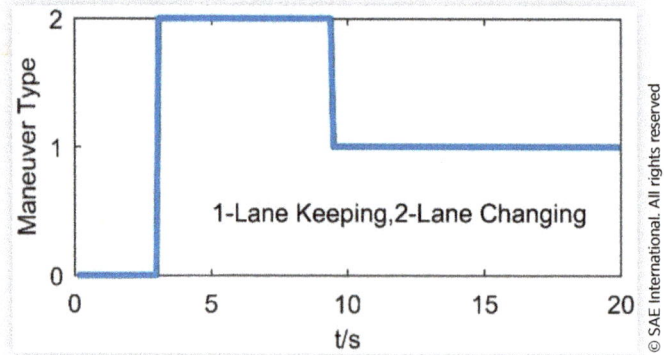

FIGURE 22 TTC calculation results.

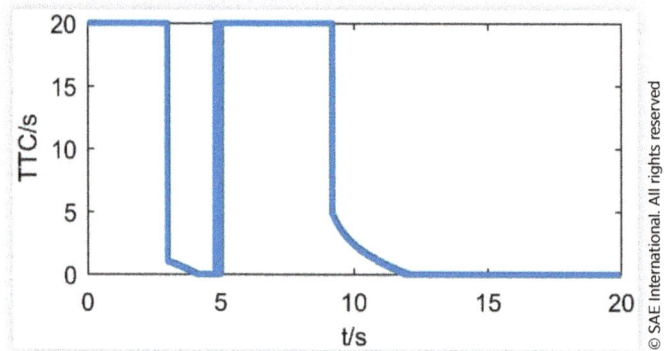

FIGURE 23 Collision symbol comparison.

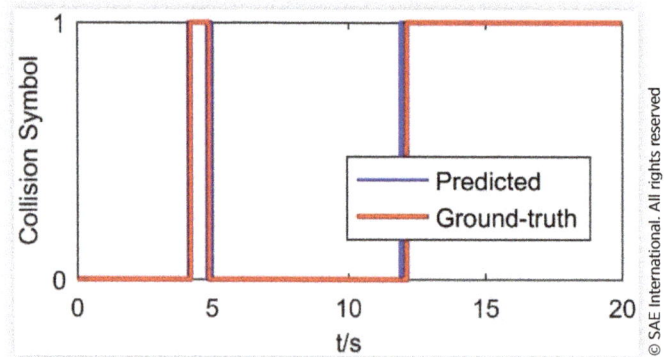

FIGURE 24 Collisions situation illustration.

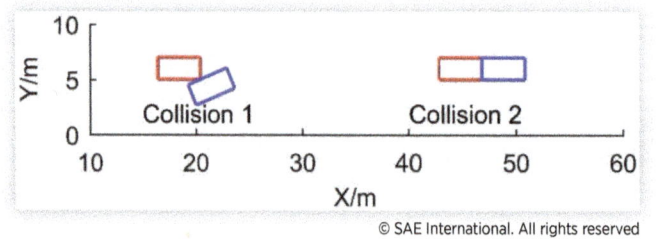

FIGURE 25 Error analysis. (Blue-collision prediction results. Green-ideal collision prediction results. Red- ground-truth collision symbol. Black-reference line.)

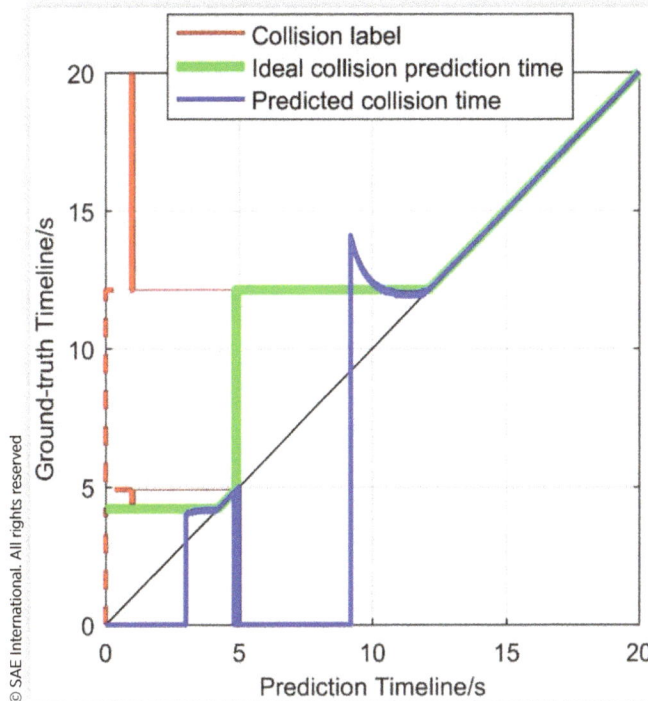

Summary

This paper describes a maneuver-based strategy to vehicle-to-vehicle collision threat assessment. First, a MTPM is built based on GMHMM. Near-future trajectories of ego vehicle and traffic vehicles are estimated with the combination of vehicle's maneuvers and kinematic models that corresponds to every maneuver. Based on the inferred maneuvers, trajectory sets consisting of vehicles' position and motion states are predicted by kinematic models. TTC is calculated in a strategy of employing collision detection at every predicted trajectory instance. Computer graphics method SAT is applied during collision prediction, so that TTC calculation can conduct with both speed and accuracy. A threat level index based on reverse TTC is used to quantize vehicle's collision threat degree. Authentic data collected in field tests are used in algorithm training, and overall strategy is validated on PanoSim. Simulation result shows that MTPM can identify maneuvers in high accuracy rate. TTC and threat level index can be calculated timely. The system error is discussed at the end of simulation.

Contact Information

For questions or to contact the authors, please email to Professor
Weiwen Deng at: kdeng@jlu.edu.cn

Acknowledgments

The authors wish to acknowledge the support of National Science Foundation of China under grant U1564211, National Key Research. Development Program under grant 2016YFB0100904 and Graduate Innovation Fund of Jilin University (Grant No.2017135).

References

1. On-road Automated Vehicle Standards Committee, "SAE J3016: Taxonomy and Definitions for Terms Related to On-Road Motor Vehicle Automated Driving Systems," SAE International, 2016.

2. Kim, B., Park, K., and Yi, K., "Probabilistic Threat Assessment with Environment Description and Rule-Based Multi-Traffic Prediction for Integrated Risk Management System," *IEEE Intelligent Transportation Systems Magazine* 9, no. 3 (2017): 8–22.

3. Eskandarian, A., ed, *Handbook of Intelligent Vehicles* (Springer, 2014).

4. Vahidi, A. and Eskandarian, A., "Research Advances in Intelligent Collision Avoidance and Adaptive Cruise Control," *IEEE Transactions on Intelligent Transportation Systems* 4, no. 3 (2003): 143-153.

5. Schreier, M.. "Bayesian Environment Representation, Prediction, and Criticality Assessment for Driver Assistance Systems," Automatisierungstechnik 65.2 (2017): 151-152.

6. Araki, H. et al. "Development of Rear-End Collision Avoidance System," *Proceedings of the 1996 IEEE, Intelligent Vehicles Symposium*, IEEE, 1996.

7. Coelingh, E., Eidehall, A., and Bengtsson, M., "Collision Warning with Full Auto Brake and Pedestrian Detection - A Practical Example of Automatic Emergency Braking," *Intelligent Transportation Systems (ITSC), 2010 13th International IEEE Conference on*, IEEE, 2010.

8. Jiménez, F., Naranjo, J.E., and Gómez, Ó., "Autonomous Collision Avoidance System Based on Accurate Knowledge of the Vehicle Surroundings," *IET Intelligent Transport Systems* 9, no. 1 (2014): 105-117.

9. Lefèvre, S., Vasquez, D., and Laugier, C., "A Survey on Motion Prediction and Risk Assessment for Intelligent Vehicles," *Robomech Journal* 1, no. 1 (2014): 1.

10. Brannstrom, M., Coelingh, E., and Sjoberg, J., "Model-Based Threat Assessment for Avoiding Arbitrary Vehicle Collisions," *IEEE Transactions on Intelligent Transportation Systems* 11, no. 3 (2010): 658-669.

11. de Campos, G.R. et al., "Collision Avoidance at Intersections: A Probabilistic Threat-Assessment and Decision-Making System for Safety Interventions," *Intelligent Transportation Systems (ITSC), 2014 IEEE 17th International Conference on IEEE*, 2014.

12. Kaempchen, N., Schiele, B., and Dietmayer, K., "Situation Assessment of an Autonomous Emergency Brake for Arbitrary Vehicle-to-Vehicle Collision Scenarios," *IEEE Transactions on Intelligent Transportation Systems* 10, no. 4 (2009): 678-687.

13. Kim, B., Park, K., and Yi, K., "Probabilistic Threat Assessment with Environment Description and Rule-Based Multi-Traffic Prediction for Integrated Risk Management System," *IEEE Intelligent Transportation Systems Magazine* 9, no. 3 (2017): 8-22.

14. Houénou, A., Bonnifait, P., and Cherfaoui, V., "Risk Assessment for Collision Avoidance Systems," *Intelligent Transportation Systems (ITSC), 2014 IEEE 17th International Conference on IEEE*, 2014.

15. Schreier, M., Willert, V., and Adamy, J., "An Integrated Approach to Maneuver-Based Trajectory Prediction and Criticality Assessment in Arbitrary Road Environments," *IEEE Transactions on Intelligent Transportation Systems* 17, no. 10 (2016): 2751-2766.

16. Russell, S., Norvig, P., and Intelligence, A., "A Modern Approach," *Artificial Intelligence*, Vol. 25 (Egnlewood Cliffs: Prentice-Hall, 1995), 27.

17. Firl, J., *Probabilistic Maneuver Recognition in Traffic Scenarios*, Vol. 31 (KIT Scientific Publishing, 2015).

18. Rabiner, L.R., "A Tutorial on Hidden Markov Models and Selected Applications in Speech Recognition," *Proceedings of the IEEE 77*.2 (1989): 257-286.

19. Lv, A. et al., "Recognition and Analysis on Highway Overtaking Behavior Based on Gaussian Mixture-Hidden Markov Model," *Automotive Engineering* 32, no. 7 (2010): 630-634

20. Akenine-Möller, T., Haines, E., and Hoffman, N., *Real-Time Rendering* (CRC Press, 2008).

21. Lin, L. et al., "A Research on the Collision Avoidance Strategy for Autonomous Emergency Braking System," *Automotive Engineering* 37, no. 2 (2015): 168-174.

epilogue

Juan R. Pimentel
Professor of Computer Engineering
Kettering University

In this book we have characterized the safety of automated vehicles into three categories or types: (1) traditional functional safety as defined by ISO 26262, (2) SOTIF, and (3) multi-agent safety. We have also included a set of ten papers that are representative of each category. Functional safety is the oldest of the categories, and there is a wealth of information, particularly for automotive. Although ISO 26262 does not cover automated vehicles, there are attempts to apply this standard to automated vehicles. SOTIF is in the process of being standardized through project ISO PAS 21448. It is noted that there is not much academic research on SOTIF and much of the available literature was originated by practitioners. Unlike SOTIF, there is a wealth of information on multi-agent safety, perhaps due to the fact that it is more intuitive to understand and to apply. However, there is a disconnect between multiuser safety and functional safety in that the former does not use a risk-based approach and thus it is difficult to discuss them in a uniform and integrated fashion.

Beyond the three generic types of safety outlined in this book, further developments of each type are happening in several categories that include sensors, actuator, computing, and communications. Some devices or products in the above categories are commercialized by supplier companies, and this enables a clear separation of safety responsibilities between OEMs and suppliers. It is possible and somewhat expected that some of these devices or products will be certified to a certain ASIL level by some certification authority. This would enable a designer to put together several components into a final product while meeting specific ASIL requirements. Again, as an industry, there is some experience with functional safety certification but not the same can be said of SOTIF and multi-agent safety. Therefore, there is a lot of work ahead.

Thus the approach and details about the safety of products or devices to be provided by companies or stakeholders are not the same. Indeed, the new edition of ISO 26262 standard includes a lot of upgrades, eliminating the weight limit, and will thereby expand its coverage to other vehicle categories, including heavier road cars, trucks, busses, and motorcycles. More specifically, the second edition will also include guidelines on the design and use of semiconductors in an automotive functional safety context which will help designing computing and communication devices for automated driving. What remains missing from ISO 26262:2018, however, is detail on how to handle the development of safe autonomous vehicles, more specifically SOTIF and multi-agent safety.

Accordingly, this book series is intended to cover all three aspects of safety, namely, (1) functional safety, (2) SOTIF, and (3) multiuser safety in five edited books as follows:

Book 1 (this book): Characterizing the Safety of Automated Vehicles
Book 2: Automated Vehicles: Multi-Agent Safety
Book 3: Automated Vehicles: SOTIF
Book 4: Automated Vehicles: The Role of ISO 26262
Book 5: Automated Vehicles: The Safety of Controllers, Sensors, and Actuators

Each book will contain an introductory chapter by the editor followed by ten SAE representative papers in each safety category. It is hoped that together these five edited books will contribute to the understanding and development of safety applied to the design of automated vehicles.